ZOONOSES
THE TIES THAT BIND HUMANS TO ANIMALS

GWENAËL VOURC'H, FRANÇOIS MOUTOU,
SERGE MORAND, ELSA JOURDAIN

Éditions Quæ

In French : Les Zoonoses. Ces maladies qui nous lient aux animaux.
ISBN : 978-2-7592-3270-3 © Éditions Quæ, 2021

Funding for this book and its translation was provided by CIRAD,
the INRAE Directorate for Open Science (DipSO), INRAE's Scientific Division
for Animal Health, the INRAE Centre of Clermont-Auvergne-Rhône-Alpes
and the I-SITE CAP 20-25 Project. This publication has been made available
in a digital form under a CC-BY-NC-ND 4.0 licence.

ISBN print : 978-2-7592-3653-4
ISBN PDF : 978-2-7592-3654-1
ISBN ePub : 978-2-7592-3655-8
ISSN : 2267-3032

Éditions Quæ
RD 10
78026 Versailles Cedex

www.quae.com
www.quae-open.com

Contents

INTRODUCTION

In 2020, COVID-19 shone a spotlight on zoonoses — diseases caused by pathogens that are naturally transmitted between humans and other animals. These pathogens can take the form of microorganisms, most commonly bacteria or viruses, or macroparasites, such as worms.

Since its beginnings 3.8 billion years ago, the biological world has been a web of interactions among organisms. Indeed, every living creature is, in fact, an amalgamation of other living creatures. Given this permanent web of interactions, certain microorganisms and parasites may end up in new host species, an erratic process facilitated by a range of factors. Sometimes these new relationships are beneficial. Sometimes they are catastrophic. Of the myriad interspecific exchanges taking place, very few ultimately succeed. The above should serve to remind us that all species host microorganisms. It is the nature of life. However, under certain circumstances, this otherwise banal reality can end up threatening the health of individuals and societies.

As human beings, we have complex relationships with animals. These connections differ across the world, as they are shaped by cultural practices, customs, traditions, and religious beliefs. Some animals are the objects of our affection. Others terrify us. In either case, animals are front and centre among our emotional connections to the living world. They improve our daily lives. Some provide us with joy, labour, or nourishment, while others simply share natural spaces with us. Each one of these interactions represents an opportunity for pathogen exchange. Certain pathogens are part of our evolutionary heritage because they were present in our great ape ancestors. The advent of domestication created an opportunity for frequent, routine contacts between humans and farm animals, thus favouring zoonosis transmission. Human and farm animal populations grew in tandem, reducing the relative representation of wildlife species among terrestrial vertebrates. At present, the ways in which we humans exploit the environment have increased how frequently we interact with wildlife. Simultaneously, intensive animal farming has transformed the conditions under which these interactions take place, with young and genetically homogenous animals crowded together at high densities.

In this book, we explore what is currently known about zoonoses, drawing upon multifarious examples. We seek to answer certain key questions: What are zoonoses? How are they transmitted? How do we learn to safely live with them? Are zoonoses on the rise? This book is an invitation to learn more about these diseases so that we can better protect ourselves and others. An essential part of this work is transforming how we interact with animals and the living world in general.

DEFINING ZOONOSES

The term "zoonosis" comes from the Greek roots ζῷον *(zôon)*, meaning animal, and νόσος *(nosos)*, meaning disease. As far back as the classical era, people observed that certain diseases seemed to pass from animals to humans, rabies serving as one notable example. However, it was not until the 19th century that the concepts of microbes, contagion, infection, and transmission were elucidated in their modern form, paving the way for the fields of microbiology and epidemiology. German physician and researcher Rudolph Virchow (1821–1902) coined the term zoonosis after noting parallels in a parasitic disease found in both pigs and humans: trichinellosis (see p. 88). The modern definition of a zoonosis is an infectious or parasitic disease whose microbial or parasitic agents are naturally transmitted between humans and other animals. In this book, we discuss disease transmission between humans and other vertebrates, mainly mammals and birds, using the terminology defined by the World Health Organisation (WHO). We also wish to specify that, in the context of this book, we use the phrase "naturally transmitted" to mean the opposite of "experimentally transmitted" and/or "rarely transmitted".

Zoonoses have been around for as long as humans have. The direct ancestors of the genus *Homo*, and more generally all the members of the various hominid lineages, were exposed to and/or infected by pathogens coming from other animal groups. Humans were interacting with animals long before *Homo sapiens* gained self-awareness. Anthropology has taught us that, earlier on in our evolutionary history, the boundaries between humans and other animals did not exist or were highly dynamic. They were shaped by context, region, and time period. In the mid-2010s, studies were carried out in northern Australia that explored how Hendra virus was viewed by Indigenous populations with traditional lifestyles (e.g., resembling those prior to European colonisation). The findings illustrate the great disparity in the attitudes of Australia's Indigenous *versus* settler populations towards this viral disease. The reservoirs for Hendra virus are flying foxes (genus *Pteropus*), which are large fruit bats. European settlers destroyed tropical forests and planted orchards, which has brought fruit bats into increasingly frequent contact with humans in inhabited areas. Hendra virus infections in humans seem to have arisen from infections in horses, animals brought to Australia from Europe. Indigenous Australians espouse certain practices when hunting flying foxes and view these animals as beneficial for the environment. Even though the pathogen has long existed in Australia, Indigenous populations have never experienced any Hendra virus outbreaks.

CAUSES OF ZOONOSES

Zoonoses are caused by pathogens transmitted between humans and animals. These pathogens may be microorganisms invisible to the naked eye, such as bacteria, viruses, tiny fungi, protozoa, or prions. They may be macroparasites, such as helminths or parasitic arthropods (see Figure 1). While we have been using the term pathogens, it would be more accurate to say potential pathogens. These species only become pathogenic under certain conditions, in certain species, and in certain individuals. Pathogenicity arises from interactions between the potential pathogen and its host (i.e., the individual that has been infected).

In fact, microorganisms are an integral part of the environment. They occur on and in our bodies. The vast majority of microorganisms do not cause sickness. Quite the opposite — they often help ensure that our bodies are functioning properly. Such is the case for our microbiota, the symbiotic or commensal microorganisms that make up the normal flora living in our intestines or on our skin, for example. It is worth noting that, since the 2000s, researchers have identified a few animal species, notably arthropods, that have few to no microbiota. In contrast, the human digestive tract houses around one trillion microorganisms, which are involved in tasks such as digestion and immunity. This figure is two to ten times greater than the number of cells making up the human body. Healthy adult humans may also harbour more than three trillion viruses, mostly bacteriophages that infect bacteria found in the intestines and mucous membranes. Furthermore, the human genome contains endoviruses, or endogenous retroviruses, which have been making themselves at home in our DNA for more than 30 million years. Their sequences represent around 8% of our genome. Generally, they are not pathogenic in humans, and some sequences have even brought us benefits. Such is the case for the genetic material contributed by the HERV-W virus, whose products are involved in physiological mechanisms and promote placenta formation.

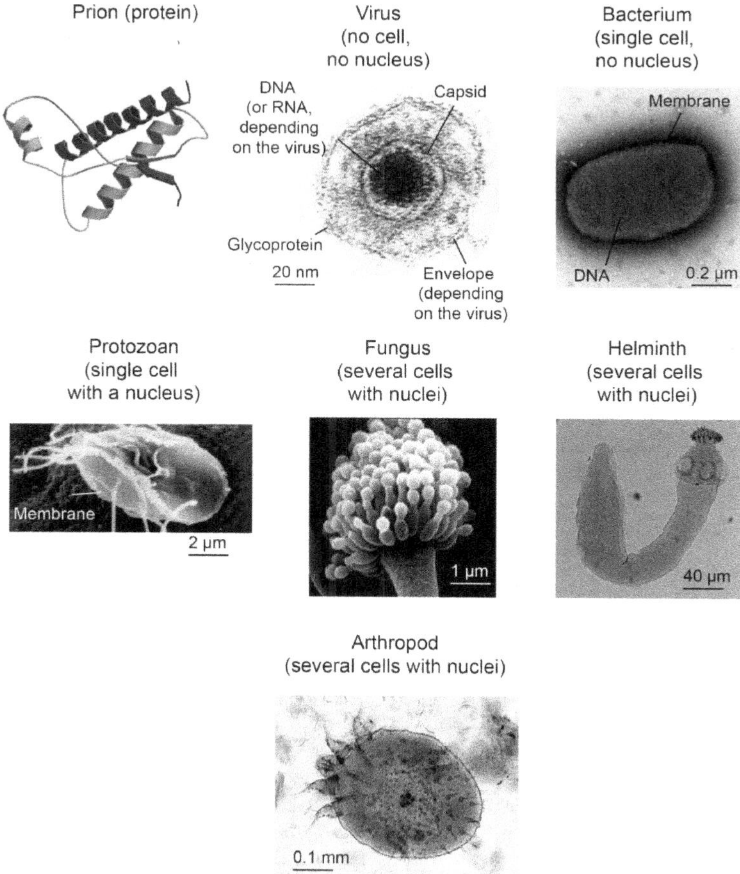

Figure 1. Examples of different zoonotic agents.

Prions, viruses (HHV-6 © Bernard Kramarsky), bacteria (*Salmonella enteritidis* © Philippe Velge/INRAE), protozoa (*Giardia intestinalis* © NIH), fungi (*Aspergillus fumigatus* © NIH), helminths (*Echinococcus multilocularis* © VetAgro Sup - Parasitology Laboratory), *and arthropods (Sarcoptes scabiei* © VetAgro Sup - Parasitology Laboratory).

NAMING CONVENTIONS FOR INFECTIOUS DISEASES

Historically, it was common to name new infectious diseases for the places they had originally been identified. For example, Crimean-Congo haemorrhagic fever was initially observed in Crimea, and its viral pathogen was isolated in the Republic of Congo. The first reported instance of Lyme disease came from the town of Lyme, Connecticut, USA. West Nile fever is caused by a virus that was isolated in the West Nile region of Uganda. However, this naming system does not reflect any sort of epidemiological reality, nor does it necessarily express accurate geographical origins. For example, the Spanish flu of 1918 was so named because Spain was the first country to publicly acknowledge the disease's existence, even though it seemingly was also present in the US. Many diseases are also named after their etiological pathogens (e.g., tuberculosis or toxoplasmosis), which may, in turn, be named after the people responsible for their discovery. Finally, some diseases are named for the animal source of transmission to humans (e.g., swine flu). However, in 2015, the WHO released recommended naming practices that do not stigmatise peoples, nations, geographical areas, and/or species. This work to change the nomenclature of emerging infectious diseases was carried out in close collaboration with the World Organisation for Animal Health (whose acronym, OIE, refers to the group's original name, *l'Office international des épizooties*); the United Nations Food and Agriculture Organisation (FAO); and the experts behind the International Classification of Diseases (ICD) tool. A disease's official name is now ultimately chosen by the ICD. Thus, the "swine flu" or "Mexican flu" that appeared in Mexico in 2009 was officially named influenza A(H1N1)pdm09.

In a 2001 study, Louise Taylor and colleagues estimated that bacteria represent one-third of zoonotic pathogens in humans. Bacteria are single-celled organisms that measure around one micrometre (µm), which equals one thousandth of a millimetre. They possess a single chromosome composed of DNA that is not contained within a nucleus. Bacteria also display a characteristic cell wall. In general, they are autonomous organisms that reproduce by binary fission. This process can be extremely fast, on the order of one division every 30 minutes. Bacteria are omnipresent within the environment, but only a small fraction of them are pathogenic. One well-known infectious agent is the bacterium responsible for tuberculosis (see p. 80). Bacteria can normally be treated with antibiotics (but see p. 82).

Parasitic worms (i.e., helminths) are thought to represent another third of zoonotic pathogens in humans. This group includes the round worms, also known as nematodes (e.g., *Trichinella spiralis*), and the flat worms, alternatively called cestodes (e.g., tapeworms and trematodes, such as the blood flukes). Their adult stages tend to be visible to the naked eye. Helminths are generally found in the digestive system, the blood, and various other tissues. Some have complex transmission cycles involving multiple host species. Parasitic worms can be treated with anthelmintics. The compounds used specifically to eliminate gastrointestinal worms are called vermifuges or vermicides. Depending on the helminth, humans may act as definitive hosts (i.e., harbour reproductive adults), intermediate hosts (i.e., harbour larvae), or dead-end hosts (i.e., do not transmit the parasite).

Viruses appear to account for one sixth of zoonotic pathogens. Generally extremely small in size (< 0.1 μm), they are composed of nucleic acids (DNA or RNA, which convey genetic information) surrounded by a protein shell called a capsid. Enveloped viruses sport an additional outer wrapping composed of lipids. They are obligate parasites that must infect cells to replicate, which ultimately disrupts normal host functioning. The rabies virus is an emblematic example of a zoonotic virus (see p. 103). Although viruses can sometimes be treated using antiviral drugs, which block the replication cycle, control strategies largely rely on shutting down transmission chains and, when possible, vaccinating populations.

Microscopic fungi are thought to account for 10% of zoonotic pathogens in humans. Like all other fungi, they sport cell walls and can spread via spores. They display a variety of lifestyles: they can grow on decomposing organic matter, live in symbiosis with other organisms, or form part of the digestive, skin, or genital flora found in humans and other animals. Fungi such as ringworm or aspergillosis can be pathogenic, especially in immu-nocompromised people, who may experience infections on their skin, in their mucous membranes, or in other tissues. Fungal infections are treated with compounds called antifungals. While this book does not discuss zoonotic fungi in great detail, a few examples are mentioned in the section on contact transmission.

CHARACTERISTICS OF ZOONOTIC VIRUSES

It is more common for zoonotic viruses to have an RNA than a DNA genome because RNA accumulates uncorrected replication errors, which can serve as fodder for evolution. RNA viruses also tend to replicate within the cytoplasm of host cells; they do not need to enter the nucleus. As a consequence, they must only make it past the cell membrane, a trait that enhances their ability to infect mul-tiple species. As underscored by epidemiologist Mark Woolhouse, the vast majority of new viruses with epidemic potential in humans are related to, but not directly descended from, other viruses capa-ble of spreading within human populations.

Sometimes, the genetic differences between zoonotic and non-zoonotic pathogens are quite small. For example, severe acute res-piratory syndrome coronavirus 1 (SARS-CoV-1) has a 29-nucleotide deletion that is absent from a closely related coronavirus found in masked palm civets (*Paguma larvata*); the former virus is pathogenic in humans, but the latter is not. Thus, while genomic comparisons can provide hints about a virus' zoonotic potential, they do not indi-cate whether or not emergence is likely.

Protozoa are estimated to represent about 5% of zoonotic patho-gens. Unlike bacteria, protozoa are complex single-celled organ-isms whose DNA is organised into chromosomes and contained in a nucleus. They vary in size from one micrometre to one millimetre. They occur in soils and aquatic environments, and only a small percentage can cause disease in humans and other animals. That said, some are obligate parasites. Given the meta-bolic similarities between protozoa and vertebrates, compounds that could be used to treat protozoa also tend to have harmful effects on their hosts. Consequently, only a limited arsenal of drugs can effectively be deployed against protozoa. Examples of zoonotic diseases caused by protozoa include toxoplasmosis (see p. 87); leishmaniasis (see p. 62), vectored by small, blood-sucking phlebotomine sand flies; sleeping sickness (African trypanosomiasis), vectored by tsetse flies; and Chagas disease (American trypanosomiasis), vectored by kissing bugs. Malaria is also caused by a protozoan. While this disease is thought to have started off as a zoonosis, it is no longer transmitted from animals to humans. The only exceptions are the malaria pathogens *Plasmodium knowlesi* and *P. cynomolgi* in Southeast Asia (see p. 18).

The main parasitic arthropods are insects and mites that parasitise the skin (i.e., ectoparasites). Sometimes, they are simply a nuisance. However, at other times, they can cause intense itching, resulting in pronounced lesions with serious health impacts. Certain members of this group, such as mosquitoes and ticks, vector pathogenic viruses, bacteria, and protozoa. Insecticides and acaricides are the most common compounds used to control arthropod pests.

Prions are proteins with abnormal spatial configurations or folding patterns. They mainly occur in the brains and spinal cords of adult mammals. Unlike viruses, bacteria, and parasites, they do not provoke infection by expressing information contained in DNA or RNA. Because of their abnormal configurations, prions are impervious to enzymatic degradation and can induce abnormal folding in normal proteins. Nervous tissue containing large quantities of prions has a sponge-like appearance, which is why prion-mediated neurodegenerative diseases in humans and other mammals are called transmissible spongiform encephalopathies. Prions are extremely resistant to conventional disinfection techniques and methods of protein inactivation. As a result, there is currently no treatment for prion diseases. To date, the only known prion-provoked zoonosis is bovine spongiform encephalopathy (see p. 105). Other prion diseases exist but are either specific to humans (Gerstmann-Sträussler-Scheinker syndrome, or kuru) or to other mammalian species (e.g., scrapie in sheep).

NON-INFECTIOUS DISEASES CAUSED BY ANIMALS

Animals can cause human illnesses via agents other than pathogens. However, such diseases are not zoonoses. For example, animals produce allergens, which provoke reactions in around 3% of the French population. Allergies to dogs and cats might affect up to 20% of the world population. People commonly react to an allergen in cat saliva, which cats spread across their fur during grooming. Horses also produce allergens, which are found in their hair, dander, and urine. Rodents can provoke extremely serious allergic reactions, especially via allergens present in their urine. In birds, the best-known allergens are found in droppings. Some of these allergens are highly volatile and can travel long distances before they end up being inhaled. Arthropods can also provoke allergies in a variety of ways: via bites, stings, or inhalation (e.g., dust mites). Additionally, animals can transfer the genes of the pathogenic or non-pathogenic organisms that they host. One example is drug resistance genes, which can be bidirectionally transmitted between humans and other animals. Bacteria tend to utilise a ubiquitous set of resistance mechanisms. More than sixty years of unrestricted antibiotics usage strongly selected for resistance in bacteria. The prevalence of antibiotic resistance is alarming and underscores the need to greatly decrease the use of these compounds in humans and other animals (see p. 82). Consequently, this topic is increasingly a part of discussions centred on zoonoses.

DISEASE RESERVOIRS

A disease reservoir is an ecological system in which a pathogen is perpetually maintained and from which it can spread into a "target" population, such as into humans in the case of zoonoses. Reservoir structure can be simple or complex. For example, the reservoir may be a population of a single animal species or a community comprising populations of various species, with each making a different contribution to pathogen transmission. There may also be an environmental component to the disease reservoir. Different reservoir hosts may vary in their susceptibility to infection.

For example, the red fox *(Vulpes vulpes)* is the only flightless mammal to serve as a reservoir for rabies virus in Western Europe (see p. 103). In contrast, an array of species act as the reservoir for Lyme bacteria, which circulate among rodent, bird, and tick populations (see p. 74). Another illustration can be seen in *Cryptosporidium parvum*, a protozoan parasite that causes acute gastroenteritis in humans. Its reservoirs are communities of numerous mammal species that occur within excrement-contaminated environments.

Ultimately, a population's capacity to act as a reservoir at a given location depends on its ability to maintain pathogen transmission on its own (i.e., intraspecifically) and successfully pass the pathogen to other host species. When pathogens are transmitted by vectors, competence can be quantified using the percentage of vectors that become infected after feeding on an infected animal. It is important to note that the above capacity is also shaped by epidemiological and ecological conditions, including factors such as population density, contact frequency, the surrounding biological community, and environmental characteristics. For example, several non-human primate species are reservoirs for chikungunya and dengue viruses because they are fed upon by *Aedes sylvestris* mosquitoes. However, in places where non-human primates are absent (other than in zoos), such as in urban areas or on Réunion Island, transmission occurs directly between humans and mosquitoes (e.g., *Aedes albopictus*).

A question naturally arises: are certain taxonomic groups more likely to act as reservoirs for zoonoses? If so, do they display particular characteristics? For example, for humans and non-human primates, it has been established that greater phylogenetic relatedness favours pathogen exchange. Similarly, there have been opportunities for pathogen exchange among humans and the species with which they have long cohabited (e.g., domestic or commensal animals). Indeed, species with a longer history of domestication share a greater number of pathogens with humans (see p. 39).

It is possible that certain taxonomic groups, such as rodents or bats, could be better reservoirs because of their life-history traits (e.g., number of offspring or lifespan); their ecology (e.g., habitat preference, gregarious *versus* solitary lifestyle, or position within the food web); their immune systems; or their physiology. This issue is still being explored (see sidebar p. 70). The number of viruses found in the different mammalian orders largely seems to correlate with order species richness but also with relative research intensity. For instance, rodents (Rodentia: 2,552 species) and bats (Chiroptera: 1,386 species) harbour significantly more viruses than do carnivores (Carnivora: 305 species). It also appears that the percentage of viruses that are zoonotic is consistent across taxonomic groups, accounting for factors such as phylogenetic relatedness (i.e., primates) and a history of cohabitation with humans (i.e., domestic animals). Thus, rodents and bats might be expected to host the highest numbers of zoonotic viruses. In addition, it may be that certain taxa are overrepresented in available data because of historical research interests.

INFECTION RISKS ASSOCIATED WITH HEALTHY ANIMALS

Zoonotic agents may be pathogenic to humans without being pathogenic to animals. For example, commensal flora in animals can cause disease when transmitted to humans. Take the case of *Pasteurella* bacteria, which occur asymptomatically in the upper aerodigestive tracts of most cats. After being bitten or scratched by cats, humans may experience local bacterial infections that must often be treated with antibiotics. Animals can also display a high level of pathogen tolerance: even when infected, hosts may not show any symptoms. Such is often seen in reservoir species, as illustrated by the rodent and bird reservoirs of Lyme disease or the bat reservoirs of various emerging viral diseases. Intestinal parasites also commonly go unnoticed in animals (e.g., roundworm infestations in dogs and cats), but can cause health issues if ingested by humans. Finally, animals may be in the incubation phase of a disease—contagious but not yet symptomatic. In short, it seems best to avoid handling unfamiliar animals, even more so if the species is wild and conditions are not conducive to ensuring health and safety. With such in mind, there is no reason to expect danger around every corner. The risk of becoming infected with a zoonosis is very low if you are interacting with familiar, asymptomatic animals kept under healthy living conditions and you are not immunocompromised.

FROM ONE SPECIES TO ANOTHER: HOST SPECIFICITY, SPECIES JUMPS, BARRIERS, FILTERS, AND OTHER CONCERNS

Compared to predator-prey interactions, host-pathogen interactions can be long lasting and quite intimate. The adjective "sustainable" has even been used on occasion. Some pathogens have coevolved with their hosts over millions of years. One result of these interactions is that pathogens may end up utilising the range of transmission possibilities available to them. Specific terminology has been developed to describe different scenarios.

Host specificity refers to the set of species that a given pathogen can infect. Thus, a generalist pathogen can infect numerous host species, while a specialist pathogen can infect a more limited number of species. A higher degree of host specificity is favoured in environments containing smaller numbers of species with high population densities. Consequently, pathogens can more effectively "utilise" the fewer species available to them. More generalist pathogens are less dependent on a given resource (i.e., host). Alternatively, we now know that different strains of a given pathogen species may be adapted to specific animal hosts, a discovery that has come about thanks to advances in genomics, which have improved our ability to characterise intraspecific pathogen diversity.

"Species jumps" describe situations in which pathogens move from one host species to another. This term is mostly employed when a pathogen has recently been detected in a new species or when such situations come as a surprise to epidemiologists, who sometimes use the questionable phrasing "crossing the species barrier". It is challenging to quantify the frequency of such "jumps". When we examine pathogen transmission patterns (see p. 35), we only see successful transmission events. Indeed, we will never know how many unsuccessful transmission events or asymptomatic transmission events have occurred because, by their very nature, such incidents slip past unnoticed. Life is woven from both continuous and discrete phenomena. An illustration is "mad cow" disease in the UK (see p. 105), which was transmitted by the consumption of contaminated meat. It was immediately apparent that meat-eating species were differentially affected. Only felids, notably domestic cats, displayed signs of illness. No domestic dogs, or indeed any other canids, showed any symptoms of the disease, even if both animal groups likely experienced the same degree of exposure.

A small-scale species jump can be seen in the history of the *Plasmodium* species occurring in humans and some non-human primates. Members of the genus *Plasmodium* cause malaria, a mosquito-vectored disease. To date, four species have been found to provoke malaria in humans: *P. falciparum*, *P. malariae*, *P. ovale*, and *P. vivax*. The origin of *P. falciparum*, the deadliest of the four, remains shrouded in mystery. The parasite seems poorly adapted to life in humans given its high level of virulence. Phylogenetics research from the 2010s suggests *P. falciparum* recently evolved from a parasite found in gorillas (*Gorilla gorilla*). The parasite is still present in gorillas but is no longer zoonotic. Instead, it served as the ancestor for a *Plasmodium* species that became a human pathogen. Other research from the 2010s described a new *Plasmodium* species responsible for human illness in tropical Asia: *P. knowlesi*. This parasite had already been observed in local macaque populations. Was this finding evidence of a species jump? Of the parasite's recent adaptation to a new host? It seems more likely to be the result of a shift in diagnostic methods: genetic tests replacing the use of light microscopy. Since then, researchers have come to realise that *P. knowlesi* was regularly confused with *P. malariae* because of their morphological similarities. It now seems that *P. knowlesi* was never actually transmitted among humans. The same morphological confusion was observed between *P. vivax* and *P. cynomolgi*, another *Plasmodium* species found in simians. Ultimately, *P. knowlesi* and *P. cynomolgi* should be viewed as macaque parasites with zoonotic potential. However, there is no indication that either is becoming a human pathogen.

People often evoke the concept of the "species barrier" in the context of species jumps, notably in research on emerging zoonoses. A "species barrier" expresses the notion of a hurdle that impedes a pathogen from moving from an established host species to a new host species. Such barriers are seen as specifically protecting humans. However, the "species barrier" is antithetical to the nature of zoonoses. In general, zoonoses and diseases affecting multiple species demonstrate that the "species barrier" can be crossed, straddled, and skirted. Then again, to successfully infect several species, pathogens must make it past what parasitologist Claude Combes has termed "filters" (see Figure 2). The encounter filter expresses the likelihood of an encounter between a given pathogen and host species; its characteristics are shaped by the ecology and behaviour of both. For example, the lancet liver fluke (*Dicrocoelium dendriticum*) is a flatworm (Phylum Platyhelminthes) that requires three hosts to complete its life cycle: a snail, an ant, and a ruminant. Infections with this species are common in animals other than humans. The latter rarely become infected, given that transmission requires the ingestion of a parasitised ant. While such a scenario is unlikely, humans have become infected by other types of parasitic worms after accepting bets to swallow slugs. Such bets were perhaps inspired by tuberculosis treatments in sanatoriums in the 19th century, which sometimes involved slug consumption. There is also the compatibility filter, which expresses the likelihood of a given pathogen bypassing a host's defence mechanisms, allowing establishment and reproduction within the host's body. Interestingly, pathogens may interact with hosts in such a way as to promote behaviours that encourage transmission. For example, ants infected with the lancet liver fluke will remain unmoving at the top of grass blades, which increases their chances of being eaten by ruminants. Similarly, mice infected with the pathogen responsible for toxoplasmosis (*Toxoplasma gondii*) show less fear in the presence of cats, which makes them easier prey.

For pathogens to move from one species to another, they must navigate processes that are ecological (e.g., that affect encounter probabilities); cellular and molecular (e.g., that affect mechanistic interactions between pathogen and host); and evolutionary (e.g., that affect genetic diversity and adaptation).

Animal and zoonotic diseases can be classified based on the relative role played by animals in transmitting the underlying pathogens to humans (see Table 1). This classification system is rather qualitative, and the distinctions between classes remain

fluid. However, it is nonetheless useful for differentiating among different epidemiological situations with regards to transmission patterns, host-pathogen interactions, and disease prevention and control. In class 1 are diseases that remain exclusively in animals other than humans. They are thus strictly animal diseases, not zoonoses. In class 2 are diseases for which transmission occurs exclusively from animals to humans. There is no transmission among humans. In class 3 are diseases for which transmission occurs mostly from animals to humans, but there may also be some human-to-human transmission. The infectious agent is not yet well adapted to the latter. In class 4 are diseases for which transmission mostly occurs among humans, but where animals can still serve as sources of infection under certain circumstances. Finally, class 5 contains diseases for which transmission occurs solely among humans even if the pathogen originated in animals. Such diseases are no longer zoonoses according to the strictest definition of the word. They are, however, of zoonotic origin.

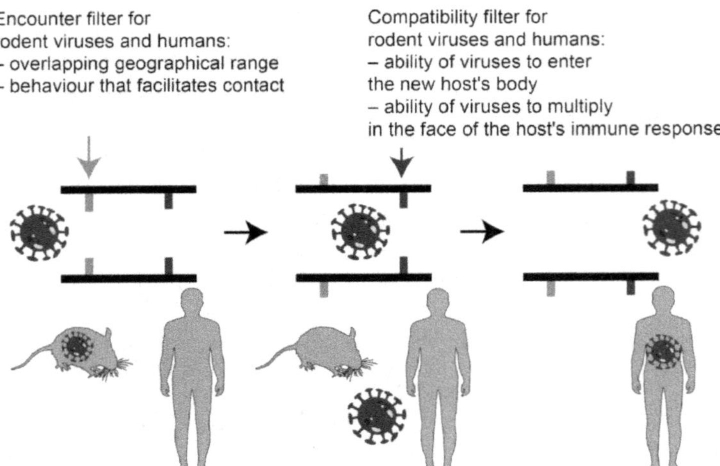

Figure 2. Encounter and compatibility filters. Here, we present the example of a virus moving from rodents to humans.

Where does COVID-19 fall in this classification scheme? The virus is of animal origin. It seems likely that the following scenario occurred: the virus was transmitted from animals to humans and then adapted to humans, allowing it to circulate without any further transmission from animals. COVID-19 could therefore reasonably be placed in class 5, even if there have been rare cases of pets, namely cats and dogs, becoming infected because they live in enclosed spaces, such as apartments, with sick, infectious humans. Transmission to captive wild species, including felines in zoos and minks on farms, has also been observed. Additionally, reverse contamination of humans working on mink farms has also been described on a few rare occasions. It came about because of the extremely high levels of airborne virus in buildings containing several thousand animals. An unusual scenario occurred in North America: white-tailed deer (*Odocoileus virginianus*) became infected as a result of living in close proximity to human beings. In early 2022 in Hong Kong, there were two cases of humans being infected by their pet golden hamsters (*Mesocricetus auratus*) (see p. 89).

Table 1. Classification of infectious diseases in animals and humans according to the relative role played by animals in transmission to humans; an example is given for each class.

Class	Transmission			Examples			
	From animals to humans	Among humans?	Zoonotic?	Pathogen	Animal reservoirs	Symptoms in animals?	Human disease
1	No	No	Strictly occurs in animals	ASF virus	Wild boars	Yes	None
2	Yes	No	Zoonotic	Rabies virus	Dogs	Yes	Rabies
3	Yes	Limited	Zoonotic	MERS-CoV	Dromedaries	No	MERS
4	Yes	Yes	Zoonotic	Yellow fever virus	Monkeys	Yes[3]	Yellow fever
5	Yes, originally	Yes	Of zoonotic origin[1]	SIV → HIV[2]	Monkeys	Yes[3]	AIDS

ASF: African swine fever; MERS-CoV: Middle East respiratory syndrome coronavirus; SIV: simian immunodeficiency virus; HIV: human immunodeficiency virus; AIDS: acquired immune deficiency syndrome.

[1] Strictly human disease of proven or probable zoonotic origin
[2] HIV (humans) evolved from SIVs (monkeys)
[3] Symptoms can be more or less pronounced depending on the simian species; for example, yellow fever is symptomatic in monkeys in the Americas but not in monkeys in Africa

FROM EXPOSURE TO INFECTION

Exposure and Pathogenicity

Our body has natural barriers that block pathogen invasions. These include the skin (provided it is undamaged) as well as the mucous membranes, whose exterior is composed of tightly packed epithelial cells and connective tissue, itself formed of cells and fibres. In the oral cavity as well as the digestive, reproductive, and respiratory systems, the mucous membranes have glands that produce mucus, a viscous substance that traps pathogens. In the nasal cavity and respiratory system, the mucous membranes have cilia or hairs that sweep away particles. There is also a series of effective barriers in the stomach and digestive tract, namely the presence of gastric acid or the resident gut flora. Another organ with a protective function is the urethra: the duct that transports urine outside the body. When the bladder empties, any bacteria or viruses inside the bladder are eliminated. Finally, coughing up phlegm helps eliminate a large number of pathogens.

However, sometimes pathogens manage to enter the body because they are helped along, by a tick or a mosquito for example. They may also exploit weaknesses, such as a cut in the skin, or they might simply manage to get past the protective barriers in other ways. Thus, some viruses or bacteria enter the body via the mucous membranes, notably by adhering to receptors on surface epithelial cells. Such is the case for influenza viruses.

After a human is exposed to a potential pathogen of animal origin, the pathogen can either be eliminated or cause an infection of varying severity. Thus, exposure does not automatically result in infection or illness. In the time before Louis Pasteur developed the rabies vaccine, accounts show that only about half of the people bitten by rabid dogs became ill and died.

If humans represent a novel host species, repeated contact will facilitate the pathogen's adaptation to the new set of environmental conditions represented by the human body. Adaptation results from changes to the pathogen's genome, via mechanisms such as mutation, recombination, and reassortment, as observed in influenza viruses.

VIRULENCE AND TRANSMISSION

The concept of virulence is strongly tied to that of pathogenicity. It refers to the degree of pathogenicity exhibited by an infectious agent—its ability to multiply within the host's body and cause pathological disease. It is used in ecology to convey the degree to which host survival or reproduction is affected by infection. The stronger the decrease, the greater the virulence. Virulence is subject to natural selection. It may seem paradoxical, but a pathogen does need hosts to survive. However, some pathogens have high case fatality rates (i.e., the number of infected people who die divided by the total number of people confirmed to be infected). Ebola, for example, has a 50% case fatality rate. Yet, if a pathogen causes too much harm to its host, its transmission will suffer as a result. In 1982, two well-known scientists, Roy Anderson and Robert May, suggested there is an evolutionary trade-off between virulence and transmission. Two years later, Paul Ewald posited that virulence depended on interaction type: if a pathogen spares its host, it can better infect other hosts. For instance, in the case of vertical transmission (from mother to child), pathogens have a "vested interest" in allowing the mother to stay relatively healthy so they can be transmitted to her offspring. Conversely, mosquito-vectored pathogens have no such need. A mosquito can easily bite a sick and feverish human who is bed bound. However, it is again important to underscore that real-life dynamics are complex. Host-pathogen relationships are influenced by multiple parameters, such as simultaneous infections by several pathogens or the intensity of the host's immune response. Factors that affect virulence may have different impacts over the short and long term. For example, over the short term, antibiotic treatments help people deal with bacterial infections. However, they also cause imbalance in the microbiota. Over the long term at broader scales, antibiotic usage favours the emergence of antibiotic resistance.

Whether or not a host will show symptoms depends both on the pathogenicity and biotic environment of the infectious agent. Pathogenicity refers to an infectious agent's potential ability to cause disease in a host. It is shaped by the agent's ability to invade and multiply in different cells and tissues; release toxins, if need be; and resist host defences. The biotic environment refers to any interactions that take place with other infectious agents and with host defences. For example, the causative agent of Lyme disease, *Borrelia burgdorferi*, causes disease in humans but not in its reservoirs (e.g., voles or field mice) or its vectors (i.e., ticks). Some infectious agents are only pathogenic in immunocompromised people, as is the case for *Babesia divergens*, a tick-vectored protozoan with a bovine reservoir. It is rare for the factors involved in these interactions to be fully elucidated.

Immune Responses

When a host's body is invaded by a pathogen, an inflammatory response is provoked during which the immune system destroys the infectious agent, eliminates damaged cells, and repairs the resulting harm. White blood cells (e.g., macrophages) ingest and destroy infectious agents. This reaction is termed the innate immune response. However, some infections can overwhelm the immune system and cause white blood cell levels to drop.

When the infection proceeds, mechanisms are implemented that specifically target the infectious agent. This reaction is termed the acquired or adaptive immune response. In this case, the immune response focuses its attack on a specific antigen, which the body has previously encountered. The acquired immune response arises from the immune system's ability to learn from, adapt to, and remember past infections. A particular group of white blood cells is activated — the killer T cells — in addition to pathogen-specific antibodies. Acquired immunity to one pathogen can provide protection against another pathogen with similar antigenic properties. This phenomenon is called cross-immunity.

The damage caused by pathogens upon infection depends on their ability to multiply. Viruses enter host cells by perforating the cell membrane; by being swallowed by vesicles; or, in the case of enveloped viruses, by fusing with the cell membrane (see Figure 3). The latter mechanism involves a lock-and-key-type interaction between a viral envelope protein and a host cell receptor. However, viruses coming from animals do not always carry the right key. Once inside, the virus hijacks the host's cellular machinery to produce copies of its genome. Finally, the virus exits the cell, either by budding off from the cell membranes in vesicles or by destroying the cell.

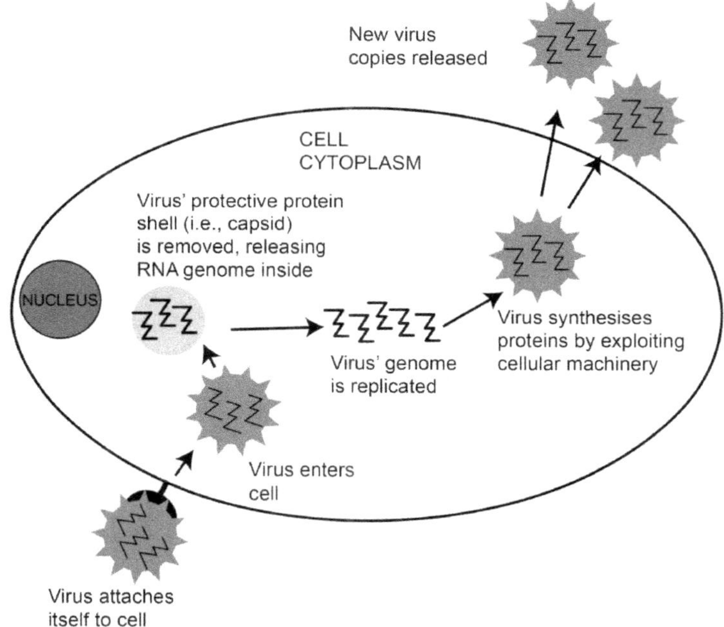

Figure 3. RNA virus replicating in the cytoplasm of a cell.

Bacteria may follow different paths. Some enter cells and hijack the host's cellular machinery in the same way as viruses. Such is the case for *Coxiella burnetii*, which causes Q fever, or *Listeria monocytogenes*, which causes listeriosis. Others are able to resist the host immune system's bactericidal activity; multiply outside the host's cells; and spread throughout the body, causing inflammation and, in some cases, secreting toxins. For example, *Escherichia coli* is naturally present in the gut flora of humans and other animals. However, some strains are pathogenic, notably enterohemorrhaegic *E. coli* (EHEC). In humans, these bacterial strains colonise the digestive tract and release toxins that damage blood vessels found locally as well as in the kidneys and brain. Finally, some bacteria can multiply either inside or outside of cells, which allows them to escape certain immune defences. They are known as facultative intracellular bacteria. An example is *Yersinia pestis*, which causes plague.

RELATIVE RISK OVER A LIFETIME

Humans are particularly vulnerable to infectious and parasitic diseases, including zoonoses, during certain periods of their lives. At greater risk are very young children, whose immune systems are not yet fully developed; the elderly, whose immune systems are less effective as a result of aging; and pregnant people, whose immune systems shift to prevent their foetuses from being expelled as foreign objects. In the latter case, one risk is that certain pathogens can cross the placenta and infect the foetus. The severity of the consequences depends on the pathogen and stage of pregnancy. There are three zoonoses worth mentioning in this context: toxoplasmosis, where primary infection can cause severe disease in the foetus; Q fever, which can result in miscarriage; and listeriosis, which can cause severe disease in the foetus and/or miscarriage.

People may also be immunocompromised for other reasons. For instance, they may have a primary immunodeficiency disorder (e.g., abnormal antibody production) or a secondary immunodeficiency disorder (e.g., resulting from chemotherapy or spleen removal). Another possibility is that they could be afflicted with a chronic illness or an autoimmune disease. Some pathogens only cause infections in the highly immunocompromised. Such is the case for *Babesia divergens*, which causes bovine babesiosis.

The immune system's ability to cope with infections is also influenced by factors that impact immune cell composition, such as smoking or stress. In general, a person's immune function and state of health are closely linked: if the body is already weak, then the immune system will provide less protection against pathogens, regardless of whether or not they are zoonotic.

DIAGNOSIS AND SCREENING

Sometimes disease diagnosis can rely on clinical signs, such as the presence of a tell-tale rash, erythema migrans, in Lyme disease infections (see p. 74). In other cases, diagnosis requires testing. Here, a test is a procedure that determines whether one or more individuals are positive, negative, or potentially positive for a given disease or infection.

Often, testing involves biological analyses and can take two forms. Diagnostic testing is performed to determine the source of health problems. Screening is used when there are no apparent health issues. The tests carried out for zoonotic diseases are the same as those used for infectious and parasitic diseases in general. They may seek direct evidence of pathogen presence, or they may look for indirect indicators of past presence, namely antibodies elicited by infection. The first type of testing determines whether the pathogen is in the sample. The second type of testing determines whether the individual has been in contact with the pathogen.

Testing is generally performed on biological samples (e.g., blood or swabs). Once the sample has been taken, it must often undergo preliminary processing (e.g., centrifugation or DNA/RNA extraction). Results are rarely available immediately given that testing procedures have multiple steps and require time, labour, and specialised equipment.

Molecular tests directly determine pathogen presence by identifying whether specific antigens or nucleic acid sequences (DNA or RNA) are present. Most commonly used are polymerase chain reaction (PCR) tests. PCR is a method that amplifies any of the target pathogen's DNA or RNA if it occurs within blood or tissue samples. Other tests look for pathogen antigens via immuno-chemical techniques, culturing, microscopy, and/or stool analysis.

Antibody testing is most often performed on serum samples, which is why the term serology testing may also be used. This approach can reveal whether different antibodies are circulating in the blood during specific and/or variable time periods. One limitation of these tests is that they can yield false positives because of antigenic cross-reactivity and false negatives if insufficient time has passed since initial infection.

This type of testing has taken a unique form in disease-sniffing dogs, which can detect the odours specifically associated with certain diseases, including COVID-19. The dogs are trained to identify the infection's volatilome: the set of volatile molecules released into the air by human cells infected with SARS-CoV-2 (see p. 89).

It is essential to understand both a test's conditions of use and limitations. For example, in the case of Lyme disease, a blood test is not helpful because the pathogen will not be present. However, serology testing can reveal the presence of antibodies, indicating that the person has been infected. The same serological approach is used for toxoplasmosis in humans. In the case of other infections, such as with West Nile virus, positive serological results may indicate that the person has been infected in the past but has likely already cleared the virus. Testing at the right time is crucial. For viral diseases such as influenza, virus presence in the blood (i.e., the viraemic period) lasts just a few days. Consequently, the virus itself will only be detectable for a short period. Conversely, antibodies appear several days into an infection and persist for varying lengths of time.

From a practical standpoint, the conditions under which samples are collected, transported, and stored can impact testing results. For example, to reliably detect a virus found in the upper respiratory tract, one must sample cells, not just secretions. If not, the test may yield a false negative result. Furthermore, depending on the pathogen, it may be important to ensure cold-chain integrity to avoid sample degradation prior to analysis. This task can be challenging in some countries.

Test performance must also be sound to obtain results that can be properly interpreted. Performance is shaped by characteristics such as implementation cost, simplicity, and/or speed as well as the test's ability to reliably classify individuals (i.e., minimise the number of false positives and negatives). Test sensitivity is defined as the likelihood of obtaining a positive result when a person is actually infected and/or sick (a true positive), while test specificity is the likelihood of obtaining a negative result when a person is not actually infected and/or sick (a true negative). Test sensitivity and specificity are intrinsic qualities that are assessed using reference analyses and samples or, in their absence, via modelling approaches.

The reliability of a positive or negative result produced by a given test can be gauged using positive predictive values (PPVs) and negative predictive values (NPVs). A test's PPV is the probability that an individual with a positive result is actually infected with the target disease. A test's NPV is the probability that an individual with a negative result is actually not infected with the target disease. The PPV and NPV are affected by test sensitivity and specificity as well as by disease prevalence. When disease prevalence is high, the PPV will be high and the NPV will be low. Conversely, when disease prevalence is low, the PPV will be low and the NPV will be high. These concepts are important when it comes to disease control efforts, as the actions taken may differ depending on a test's predicted number of false positives or negatives.

COURSE OF INFECTION

Within Individuals

Once a pathogen has successfully entered a host and started to multiply, the infection can take several different courses. For some pathogens and in some individuals, the infection may never trigger any symptoms, a situation described as asymptomatic. In other cases, there is an initial incubation period, which is the time between the moment of exposure and the first symptoms of disease. The length of the incubation period varies. For example, it can last between two days and three weeks for Ebola virus, but it ranges from several weeks to, in rare instances, a year for the rabies virus. During this period, a person may or may not be infectious — that is, they may or may not be capable of spreading the pathogen to others (for zoonoses in class 3 or above). This length of time is called the infectious period. For example, in the case of Ebola, individuals are not contagious until they develop symptoms. The same was true for SARS-CoV-1 in 2002. For other viruses, like SARS-CoV-2, individuals may be contagious a few days beforehand. In such situations, it is harder to disrupt the human chain of transmission.

Infections may be acute, characterised by the sudden onset of short-term symptoms. For example, salmonellosis causes vomiting, severe diarrhoea, and fever. These symptoms appear 2–3 days after the consumption of contaminated food. Depending on the pathogen, the acute phase of the disease may end with recovery, chronic illness, or death. For example, influenza A(H5N1) can result in severe acute pneumonia, with an estimated human case fatality rate of 60%.

Chronic illnesses generally have a gradual onset and continue over the long term. As long as a person is actively infected, the pathogen will continue to multiply. For example, certain people who come down with Q fever develop a chronic infection that can affect the heart, blood vessels, and/or bones. Brucellosis is another example of a disease that can become chronic.

The infectious agent may also persist in a host's body without multiplying or manifesting itself. This phenomenon is called a latent infection. The pathogen goes into a type of stasis, situating itself in organs that cannot be easily accessed by the immune system, which makes elimination difficult. For instance, the infectious agent responsible for toxoplasmosis goes latent once acquired. However, toxoplasmosis infections, among others, can go from latent to active if the host's immune system is somehow impaired (see sidebar p. 26).

ZOONOSES AND HUMAN CANCERS

Seven viruses, one bacterium, and three parasites have been formally identified as carcinogenic hazards by the WHO. They cause more than one sixth of human cancer cases worldwide. Most are of human origin. There are two exceptions among the parasites. The first is *Opistorchis viverrini*, found in Southeast Asia, and the second is *Clonorchis sinensis*, found in East Asia. The larvae of both species are hosted by fish and can be transmitted to humans if the latter consume raw or poorly cooked fish. Infection can cause gallbladder cancer. One of the viruses is also of zoonotic origin: HIV-1. It is classified as a carcinogen because it causes immunodeficiency, which increases the risk of developing a range of cancers.

Evidence suggests other parasites may also act as carcinogens, but definitive links have yet to be established. Finally, there is growing interest in the hypothesis that animal cancers could be zoonoses. That said, few findings indicate the occurrence of animal-to-human transmission or the existence of major risks for humans.

Within Societies

A zoonotic agent can cause an epidemic once it has infected a large number of people, who form transmission clusters in a given area over a given period of time. A pandemic occurs if an epidemic spreads over several continents, like the plague historically or COVID-19 in 2020. Epidemics usually occur when rates of human-to-human transmission are high, but such is not always the case. Infections can occur via airborne transmission (e.g., the Q fever outbreak in the Netherlands in 2007–2010), foodborne transmission (e.g., mass food poisoning), or vector-borne transmission (e.g., the West Nile epidemic in the US after the virus' introduction in 1999).

Modelling can be used to assess the risk of epidemics occurring in human or animal populations. When the transmission rate among humans is high, transmissibility is quantified using the basic reproduction number (R_0). R_0 is the expected number of individuals that will be infected, on average, by an infected individual during their infectious period in a fully susceptible, uninfected population. The value of this metric differs depending on the infectious agent and the characteristics of the host population. It is not measured from epidemic curves, but rather from observations and mathematical models.

The effective reproduction number (R_e) is the number of infections that an individual is actually responsible for at a given time. It may be lower than R_0 if, for example, a certain proportion of the population is immune. When the reproduction number is greater than 1, it means numbers of infections are climbing, leading to an epidemic. When the reproduction number is less than 1, numbers of infections are declining, and the chain of transmission will eventually be broken.

The value at any given point in time depends on the frequency of contacts involving infected individuals, the probability of transmission during contact, the duration of an infection "generation" (basically equal to infectious phase length), and the size or frequency of the susceptible population. The objective is to act upon these four factors to prevent or reduce the size of an epidemic. For example, vaccination is a strategy for reducing the size of the susceptible population.

It is also worth mentioning the dispersion factor (kappa, or k), which conveys the degree of variability in the reproduction rate within a population. The value of k ranges between 0 and 1. If k is close to 1, the number of infections produced by each infected individual is approximately consistent. If it is close to 0, there is dramatic variability. Some infected individuals contribute very little to transmission, while others contribute greatly.

Thus, to get a sense of the potential impacts of various diseases, it can be useful to compare R_0 and case fatality rate for different pathogens (see Figure 4).

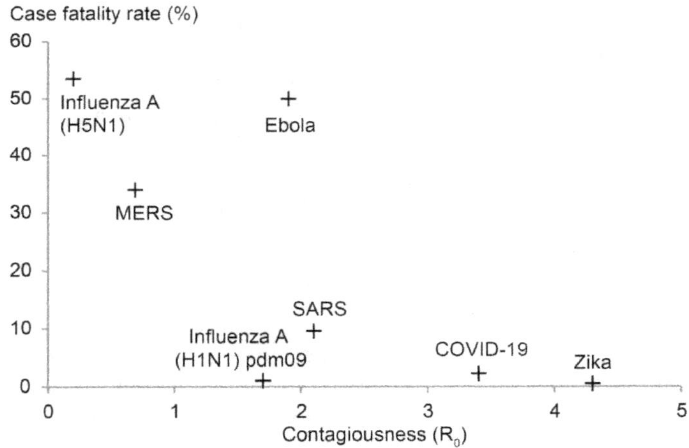

Figure 4. Basic reproduction number (R₀) and case fatality rate (number of deaths divided by the number of confirmed infections) in humans for a range of zoonotic diseases.

MERS-CoV: Middle East respiratory syndrome coronavirus; SARS: severe acute respiratory syndrome (2002–2003)

These figures were taken from the scientific literature. They should be viewed as estimates and are context dependent. They are based on various sources (doi:10.3934/mbe.2019174; doi:10.1016/S1473-3099(16)00153-5; doi:10.1016/S1473-3099(20)30484-9; doi:10.1097/EDE.0b013e3182a67448; doi:10.1056/NEJMsr1513109; https://www.who.int/news-room/fact-sheets/detail/ebola-virus-disease; https://reacting.inserm.fr/wp-content/uploads/2020/10/COREB_REACTing_13102020-compresse%CC%81.pdf; https://www.who.int/news-room/commentaries/detail/estimating-mortality-from-covid-19; doi:10.1016/j.envres.2020.109114; doi:10.1016/j.ijid.2019.08.033).

When a pathogen is primarily transmitted to humans by an animal source, it is necessary to build models that estimate transmission rates between humans and different potential animal hosts. Such multihost models must incorporate diverse parameters describing reservoir and vector population dynamics, such as human-animal contact and transmission rates as well as environmental factors.

During an epidemic, some individuals recover, while others die. In the case of many diseases, those who recover develop a degree of protective immunity. Immunity can also be acquired via vaccination, given that vaccines are available for a number of viral and bacterial diseases (see sidebar p. 115). As immunity grows, the epidemic curve flattens. The number of new infections sharply declines because a smaller percentage of the population is susceptible. This effect is called herd immunity. For more contagious pathogens (higher R_0), a greater level of population immunity (% immune individuals) is needed to limit transmission.

HISTORY AND DYNAMICS OF ZOONOSES

Life is organised into networks that span multiple scales: microorganisms, macroorganisms, populations, communities, ecosystems, and biomes (i.e., sets of ecosystems with identical ecological conditions that cover large geographical areas). Classifying biological relationships is thus challenging, especially since humans cohabit with other vertebrates, displaying a variety of overlapping relationships. Furthermore, within each species, each individual is actually a symbiosis — a vertebrate host carries around diverse microbiota, composed of viruses, bacteria, fungi, and parasites.

The other vertebrate species that share our world range from those living in nature far from humans (i.e., wild animals) to those who share our living spaces (e.g., domestic animals). There are the species that we consume, which may be wild in nature, wild but in captivity, or domesticated and raised for that specific purpose. In the category of companion animals are species that have been domesticated for several millennia as well as wild animals, which may be exotic, more or less tame, and/or not always legally obtained. For the latter, it is often the case that little is known about their biology, ethology, and probable health risks. We must not forget synanthropic species: they live in habitats close to or overlapping with ours. Their relationship with us is often commensal but sometimes parasitic.

HISTORICAL FORCES BEHIND ZOONOSES

The history of life on Earth can be represented by phylogenetic trees, on which can be seen the degree of relatedness between humans and other taxa. The human species, *Homo sapiens*, is a primate belonging to the hominid family. As a result, it shares certain traits with its extant great ape relatives, notably similar receptivity and susceptibility to certain microorganisms. It has also coevolved with a selection of shared and inherited pathogen lineages over geological time.

Phylogenetic Heritage

The earliest potential microbial inheritance received by a hominid primate seems to have been passed along to an ancestral placental mammal. Biologists have long wondered why female placental mammals manage to successfully gestate their foetuses *in utero*, without the developing offspring being expelled by the body. This ability seems to be related to the early acquisition of specific retroviruses: HERV-W and HERV-FRD. Marsupial mammals appear to be incapable of full-term intrauterine gestation. It is perhaps for this reason that their foetuses generally continue their development in the taxon's emblematic pouch, where they remain attached to their mother's teat. Consequently, these viruses could only have been encountered by an ancestor along the branch that led to the placental mammals. We can consider, in this case, that these viruses of animal origin became symbiotic over evolutionary time.

There are other examples of microorganisms whose evolutionary histories are intertwined with those of their hosts, although not always quite so intimately. *Herpesviruses* are highly host specific, with receptivity and especially susceptibility patterns that are very different depending on whether transmission is intra- or inter-specific. Good examples are provided by herpes simplex viruses in humans and herpesvirus B in Asian macaques. If a human is infected by a macaque herpesvirus, we are looking at a zoonosis. In contrast, human herpes simplex viruses were inherited from a primate ancestor over the course of the evolutionary history that gave rise to *H. sapiens*.

Within mammals can be found a single-celled, fungal parasite of the respiratory system — *Pneumocystis* — with a phylogenetic history that displays a fairly convincing degree of parallelism with that of its host species. There is a clear shared history with animals, even if new data have raised questions about the strength of this parallelism. The fungus' presence does not automatically result in disease. Instead, disease arises when the host experiences diminished immune resistance, which can occur for any number of reasons.

Lice infestations are another example of a parasitic zoonosis. Lice are haematophagous ectoparasites and carry out their entire direct life cycle in their hosts' pelage, which is not the case for fleas (also insects) or ticks (acarians). Their phylogenies partly parallel those of their hosts. However, lice species did end up jumping between primate species. In humans, these insects have adapted to the loss of body pelage by taking advantage of an artificial form of body coverage: clothing.

Hunting

The human diet has long been eclectic. Before the advent of agriculture, humans mostly consumed the diverse resources they could glean from hunting, fishing, and harvesting plants (e.g., fruits, roots, tubers, and/or leaves). Without even considering the human illnesses spurred by toxic animal and plant compounds, it seems highly likely that a vast array of microorganisms and parasites could have been transmitted to humans via the handling of prey. Epidemiological cycles are an ancient component of food webs. The arrival of a new species on the scene — *H. sapiens* in this case — allowed pre-existing cycles to expand or new cycles to progressively emerge via adaptive drift.

Humans were exposed to new microorganisms during the various actions involved in processing prey (i.e., slaughtering the animal or dressing and cutting up the carcass). The butchering process results in the exposure of people other than the hunters. For a long time, people handled dead prey with their bare hands. They still do in some cases. Human hands often have small lesions or abrasions, which can serve as entry points for microorganisms if someone touches an acutely infected animal. HIV viruses appear to have emerged because several simian immunodeficiency viruses (SIVs) were transmitted from monkey prey to human hunters in central and western Africa in the early 20th century. Such was the first phase in the evolution of SIVs to HIVs. Human behaviour then led to the epidemic in the second half of the 20th century. Although its origins were in Africa, the human species did not harbour these types of viruses, unlike other hominid species.

More recently, the SARS-CoV-1 outbreak in southern China in late 2002 appears to trace back to animal butchers (see p. 90).

Human Diets

Visible to the naked eye, macroparasites have undoubtedly had an important role in elucidating issues related to human health and illness, particularly in the context of human diets. Humans probably noted the resemblance between certain macroparasites occurring in the organs of their prey and the parasites that they themselves were infected with and shed. It leads us to wonder: what were the reactions of early hunter-gatherers? Even at present, intestinal parasites are common in a large percentage of the global human population. Many traditional treatments are still utilised. Some are likely quite ancient, as suggested by the plants found in the pouch of Ötzi, a man born 3,300 years before the common era (BCE), whose frozen, mummified corpse was discovered in an Italian glacier in 1991.

Many trematodes, nematodes, and other parasitic worms require two hosts in their life cycles. Humans may serve as either the intermediate host (i.e., for larvae) or the definitive host (i.e., for reproductive adults). In many cases, the parasite's life cycle can only be completed if the preceding host is consumed by the subsequent host. Indeed, humans were long the potential prey of various predators or scavengers. In other cases, eggs or alternative types of infectious forms are released into the environment. Hosts are infected when they ingest contaminated food.

A broad range of situations are represented in parasitic worms. Humans acquired several cestode species from their animal prey. Adults of the beef tapeworm *(Taenia saginata)* and pig tapeworm *(T. solium)* parasitise the human digestive tract. They were originally the parasites of wild ungulates, in which their larvae occur. Both secondarily adapted to humans. In turn, humans transmitted their worms to domestic bovines and pigs during the early Neolithic. Conversely, tapeworms in the genus *Echinococcus* parasitise the intestines of various carnivore species as adults, and humans act as their intermediate hosts. *Echinococcus multilocularis* and *E. granulosus* are two species found in Europe whose habitual intermediate hosts are field rodents and ungulates, respectively (see p. 85). Epidemiologically speaking, humans are almost always dead ends. However, a unique pattern is seen in a region of eastern Africa, where an ethnic group leaves out their dead for wild carnivores and scavengers, leading to a possibly closed *Echinococcus* life cycle.

In the realm of bacterial diseases, infections caused by *Helicobacter* species nicely illustrate historical transmission from humans to animals under highly specific circumstances. These bacteria cause or contribute to causing gastric ulcers. The so-called "human" *Helicobacter* species, *H. pylori*, appeared around 100,000 years ago in Africa. A 2011 study found another *Helicobacter* species, *H. acinonychis*, in the African lion (*Panthera leo*). The best explanation for this bacterium's occurrence in a felid seems to be that lions historically preyed on humans, resulting in an orally transmitted infection. Indeed, genetic analyses of the two species revealed that *H. acinonychis* diverged from *H. pylori* about 50,000 years ago.

Animal Domestication

The reasons for which humans began domesticating animals are quite complex and not fully understood. In 1967, historian William McNeill was the first to suggest that humans acquired certain infectious diseases as a direct consequence of animal domestication. This hypothesis can now be quantitatively tested using available data for zoonoses in tandem with the dates and locations of domestication events, obtained from archaeological and population genetics research. The number of infectious and parasitic diseases shared by humans and domesticated animals is indeed dependent on time since initial domestication. More importantly, the contribution made by a given species to the web of zoonoses shared with other species and humans is correlated with time since initial domestication (see Figure 5). In other words, the longer an animal has been domesticated, the more infectious agents it shares with other domesticated animals and humans. When a new species starts down the path towards domestication, it adds its pathogens to this web and becomes infected with the pathogens already in the web.

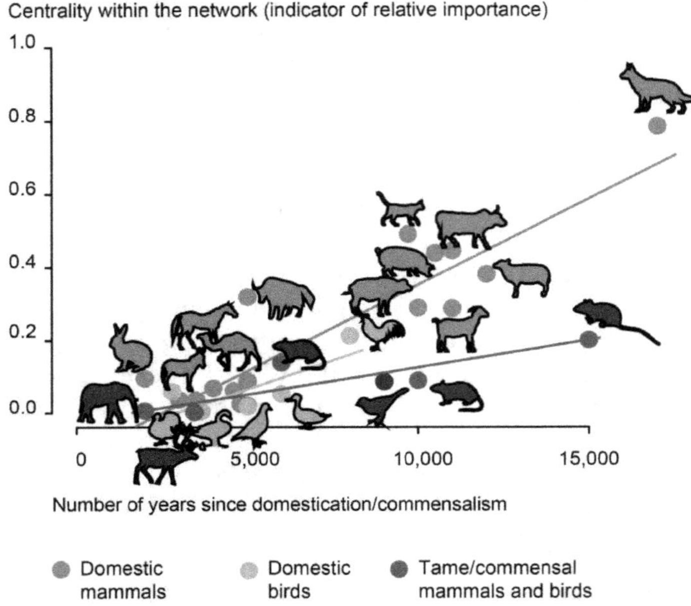

Figure 5. Animal domestication and zoonotic agents.

Relationship between a) time since domestication/commensalism in years and b) the centrality of zoonotic agents across host species within the web of agents shared by all host species (humans and other animals). Centrality is a metric indicating the relative importance of each agent. Domestic mammals: dogs (*Canis familiaris*), cats (*Felis catus*), yaks (*Bos grunniens*), zebus (*Bos indicus*), cattle (*Bos taurus*), buffalos (*Bubalus bubalis*), pigs (*Sus scrofa*), sheep (*Ovis aries*), goats (*Capra hircus*), horses (*Equus caballus*), donkeys (*Equus asinus*), dromedaries (*Camelus dromedarius*), camels (*Camelus bactrianus*), and rabbits (*Oryctolagus cuniculus*); domestic birds: ducks (*Anas platyrhynchos*), geese (*Anser anser*), chickens (*Gallus gallus*), and pigeons (*Columba livia*). Tame/commensal animals: reindeer (*Rangifer tarandus*), Asian elephants (*Elephas maximus*), Norway rats (*Rattus norvegicus*), black rats (*Rattus rattus*), house mice (*Mus musculus*), and house sparrows (*Passer domesticus*) (© S. Morand).

Also included in this group are commensal animals, such as rats, mice, and sparrows, and tame animals, such as reindeer and elephants.

Thus, historical physical proximity and, notably, this proximity's duration are better at explaining patterns of microbial exchanges than is phylogenetic proximity. Measles provides perhaps the most telling example. The now-extinct aurochs *(Bos primigenius)* was the ancestor of all domestic bovines (humped and non-humped).

Its range extended across a large swath of Eurasia and northern Africa, but it never reached the Americas. It was domesticated in at least two separate places: Mesopotamia and the Indus Valley. There is ongoing debate as to whether domestication also took place in China and at a second location in northern Africa. Rinderpest virus (genus *Morbillivirus*, family Paramyxoviridae) was "domesticated" alongside the aurochs, its host. Domestication resulted in sustained contact between the aurochs and humans, and the virus thus adapted to *H. sapiens*. This is the hypothesised origin of the measles virus, which would have emerged a few millennia ago. That the measles virus evolved in Eurasia is supported by the fact that other *Morbillivirus* species have never yet been described in other hominids to date. Furthermore, Indigenous Americans were extremely susceptible to the disease when it was introduced during European colonisation.

Fittingly, it seems that the infectious agent responsible for bovine tuberculosis (*Mycobacterium bovis*) evolved from the infectious agent responsible for human tuberculosis (*M. tuberculosis*). Thus, herders appear to have been the ones to infect their livestock, not vice versa.

Some of the health issues associated with animal domestication are clearly more recent. For example, the emergence of Middle East respiratory syndrome coronavirus (MERS-CoV) in 2012 raises many questions that have yet to be answered.

Companion Animals

Our pets may or may not be domestic animals. The dog (*Canis familiaris*) is our oldest known companion animal. It underwent domestication at least 15,000 years ago, before the beginning of the Neolithic. However, its steady companionship has come at a cost to our health. It is possible that the rabies virus has long existed in the wolf (*Canis lupus*), the ancestor to all dogs (see p. 103). Although many different wild mammal species can carry rabies, virtually all human deaths from rabies (estimated at around 50,000–60,000 per year worldwide) can be attributed to domestic dogs, most often unvaccinated or living as strays. Humans very rarely catch rabies from wild mammals.

Another "gift" from dogs to humans may have been fleas (*Pulex irritans*). Generally, fleas are only found on mammals that have a dedicated nest or burrow. Flea larvae live in the nesting material and become parasitic only after they moult for the last time and turn into adults. However, neither the Old World monkeys (i.e., cercopithecids) nor the nomadic, non-human hominids have fleas. These two groups share the commonality of often changing the sites where they bed down. At present, *H. sapiens* is the only species in these two families to be parasitised by fleas. *Pulex irritans* may have evolved from a flea occurring on commensal rodents or domestic dogs.

The shared history of humans and domestic cats (*Felis catus*) seems more uneventful. The cat likely became a companion to humans because of its behavioural tendency to hunt small rodents. This relationship has led humans to experience toxoplasmosis and cat scratch disease. Other types of zoonotic infections are much rarer.

Humans have started adopting new types of pets, sometimes called zoological companion animals (ZCAs). These are small companion animals other than cats and dogs. Little is known about the biology, behaviour, or health risks associated with certain ZCAs sold commercially and found in people's homes. There are an estimated 18 million ZCAs in France. In 2020, the most common ZCAs in Europe were small mammals (30 million), ornamental birds (52 million), and reptiles (9 million) (figures from statista.com). In general, ZCAs tend to be mammals, such as rodents (e.g., rats, mice, gerbils, chinchillas, hamsters, and guinea pigs), lagomorphs (e.g., rabbits), and carnivores (e.g., ferrets). Additionally, there are reptiles, including snakes (e.g., boas, pythons, and corn snakes), lizards (e.g., iguanas and geckos), turtles, and birds. Mygalomorphs are also represented, as are insects such as mantids. Finally, the list includes fish and amphibians. For example, some people keep axolotls (*Ambystoma mexicanum*), a type of salamander that can live its entire life and even reproduce while retaining larval traits. The axolotl's regeneration capacity is the focus of much research attention. There are laws regulating the possession of certain ZCAs by private individuals.

When ZCA species are raised in captivity, they do not tend to cause more health concerns than usual. However, there is a common yet false assumption that reptiles cannot pass along any microbes to humans. Like most vertebrates, reptiles harbour *Salmonella* species, bacteria that cause salmonellosis. Characterised by severe diarrhoea, this disease can be serious in vulnerable people. Certain ZCA species are always collected in the wild, and thus the term companion animal is truly a misnomer. When these ZCAs are imported, there is a real risk of introducing a zoonosis. Such an outcome is illustrated by the monkeypox virus outbreak that took place in the US (see p. 47). There were also four deaths in Europe between 2015 and 2016 that resulted from bornavirus infections transmitted by "pet" variegated squirrels (*Sciurus variegatoides*), a species that cannot lawfully be kept. Did the squirrels pass along pathogens to their owners, or was it the other way around? What was the original source of these viruses?

Commensal Animals

In addition to the groups of species mentioned above, there is another type of animal that lives side-by-side with humans, sometimes openly, sometimes surreptitiously. They are known as commensal animals, and representatives can be found in many different taxa. Let us envision how tempting our food reserves, granaries, and tanks of fresh water must be to certain groups of animals, especially in regions that experience seasonal dry periods. Rodents in particular have adapted to take advantage of the situation. The first were species such as the house mouse (*Mus musculus*) and the black rat (*Rattus rattus*), and they were soon followed by the Norway rat (*R. norvegicus*) and the Polynesian rat (*R. exulans*; in the Pacific). Depending on the part of the world, other species can also affect the health of the humans with whom they cohabit. The bubonic plague, caused by *Y. pestis*, is mostly spread via the black rat and its fleas, particularly *Xenopsylla cheopis* (see p. 77). The Norway rat is a strong swimmer and may play a role in spreading another bacterial disease, leptospirosis, which is transmitted by water. The Norway rat also carries a strain of hantavirus, Seoul virus, which has spread as far across the globe as the rat itself (see p. 66).

TRANSMISSION OF ZOONOSES

Pathogens make their way into our bodies using various entry points: our skin, weaknesses in our skin (e.g., wounds, bites, or stings), or our mucous membranes (e.g., in the eyes, digestive tract, and/or respiratory system). Many pathogens can use multiple approaches.

Petting and Touching Animals: Contact Transmission

Physical contact with our pets is comforting and serves as a form of communication and exchange. When we are familiar with certain animals because we are around them every day, we may forget about the diseases they are capable of transmitting. Indeed, such risks are low if the animals are healthy, well cared for, and the regular recipients of antiparasitics. However, it is always better to take precautions if you do not know where an animal comes from; if the animal is in poor health or poorly groomed; or if the animal displays aggressive behaviour. Always avoid touching wild animals.

The coats of animals may be soiled with faeces and thus covered with parasite eggs or bacteria that you could end up ingesting if you then put your fingers in your mouth. You should therefore always wash your hands before eating if you have been petting your furry companion. In addition, watch for behaviour such as intense shedding or scratching. The former might indicate the presence of ringworm, a common fungal skin infection in animals that is highly contagious to humans. Such fungal agents are found around the world. They affect many species of mammals and, on rarer occasions, bird species. Most fungal skin diseases are transmissible to humans. They often cause circular, clearly delineated rashes. These areas may display hair loss but are not itchy. Conversely, scratching can indicate that an animal has scabies, a disease caused by *Sarcoptes* mites. Although people are most often afflicted with human scabies, zoonotic infestations also occur: dog scabies can cause rashes in humans even if the mites do not breed on us.

TOUCHING DEAD ANIMALS

In general, it is best to remain cautious when you come upon dead animals and to always handle them with gloves. Infection can occur via contact with pathogens found on the animal's skin, in its secretions, and in association with any lesions. We have already brought up tularaemia. Below, we discuss erysipeloid/erysipelas and anthrax. A bacterium with a complicated name—*Erysipelothrix rhusiopathiae*—causes a disease known as erysipeloid in humans and erysipelas in animals. Erysipelas is most commonly seen in domestic pigs, sheep, and birds. People become infected after handling the viscera of dead animals. Although the disease occurs at low frequencies across the globe, 30–50% of domestic pigs are carriers of the bacterium.

Anthrax is caused by *Bacillus anthracis*. It is a disease with a global distribution that affects many animal species, especially herbivorous mammals. In Europe, anthrax cases are rare and heterogeneously distributed, with only a few cases reported each year. It is largely an occupational disease. People can become infected after handling anthrax-positive live animals, dead animals, or animal products. The bacteria or their spores can enter the body via microlesions in the skin. Transmission can also occur by means of inhalation or ingestion. Cutaneous anthrax tends to result in local infections; it is the most common form of the disease. Pulmonary and gastrointestinal anthrax are much rarer but can result in sepsis.

It is also worth noting that infection can occur via wounds to the skin or the ocular mucus membranes. Such is notably the case for tuberculosis. To limit transmission risks, professionals (e.g., veterinarians, slaughterhouse workers, and butchers) wear protective equipment when working with dead animals during autopsies, collection, or disposal, for example.

If you come across several dead wild animals or notice an animal whose death seems suspicious, you should notify a veterinarian or wildlife authorities. Europe boasts the European Wildlife Disease Association Network for Wildlife Health (https://ewda.org/ewda-network/), which was launched in Brussels in 2009.

You may have asked yourself if you could get fleas from your pets. Flea species are highly host specific. They generally live on a single species: dog fleas on dogs, cat fleas on cats, and human fleas on humans. However, if their habitual hosts are not available, fleas can feed on other species, although the infestation will come to a quick end. Fleas are often picked up by unwitting hosts who pass by a pet's bedding where non-haematophagous flea larvae are developing. After undergoing their final moult to become adults, they await their host's return. A common scenario plays out in vacation houses where a dog or cat was present the summer before. When people arrive at the house the following spring, they may be greeted by a horde of hungry fleas. It is unpleasant, but it does not last long.

Less frequently, pathogens may be transmitted by contact. Consider, for example, the bacterium *Francisella tularensis*, which causes tularaemia, a disease of varying severity. It can be found in a number of animal species. However, in Europe, the brown hare (*Lepus europaeus*) is the main host responsible for transmission to humans. Because *F. tularensis* can penetrate even healthy skin, contamination can occur after simply handling an infected animal, whether dead or alive. Tick-vectored transmission also occurs. Additionally, it is worth mentioning *Brucella* species, which cause brucellosis. This disease occurs worldwide and is responsible for around 500,000 infections in humans per year. *Brucella* bacteria are transmitted via contact with infected ruminants or pigs, mainly during events such as births or abortions, contact with products from infected animals and the ingestion of raw milk or raw milk cheese. Contact transmission also results in the spread of viruses in the family *Poxviridae*, which cause skin lesions. Contrary to what its name might suggest, cowpox virus is mainly maintained in wild rodent populations. Cats may become infected if they hunt down an infected rodent; they can then potentially transmit the virus to humans. The disease is usually mild, although it can become severe in people who are immunocompromised. Closely related viruses include buffalopox virus in India, transmitted by a milker's hands if the person has microlesions, and monkeypox virus (another misnomer, given that its reservoir is rodents, not primates), which is also transmitted by handling infected animals. A monkeypox outbreak occurred in 2003 in the US following the importation of 800 wild African rodents, including Gambian pouched rats (*Cricetomys gambianus*) and woodland dormice (*Graphiurus murinus*). After arriving in pet stores, these species passed along the virus to native North American prairie dogs (*Cynomys ludovicianus*) housed in nearby cages. The prairie dogs then transmitted the African virus to their new owners. The fact that poxviruses are quite resistant to external conditions facilitates indirect transmission.

Licking, Biting, and Scratching: Transmission via Broken Skin

Animal saliva contains many of the oral flora's commensal bacteria, including some that can cause serious infections. In animals, licking is a form of communication, but care should be taken if someone has microtears in their skin or their health is fragile.

In contrast, animal bites are never harmless. Not only do they cause pain and potentially serious tissue damage, but they can also lead to infections caused by microbes within the oral flora, such as *Pasteurella*. The result can be pronounced pain and local inflammation. For example, a rat bite can provoke a *Streptobacillus* infection. Dog or cat bites can transmit *Capnocytophaga canimorsus*, which can lead to serious illness in vulnerable individuals. The pathogen of greatest concern, however, is undoubtedly the rabies virus (see p. 103).

The most common disease associated with animal scratches is cat scratch disease (or benign lymphoreticulosis by inoculation). As its name implies, the disease is transmitted by cats; *Bartonella henselae* is the infectious agent. A more infrequent form of transmission occurs via flea bites. The main sign of infection is enlarged or swollen lymph nodes (i.e., adenopathy), a symptom that usually subsides naturally over several weeks or months.

Regardless of the situation, you should always wash and disinfect any scratches you experience.

Soilborne Transmission

The soils of the earth are extremely diverse and complex ecosystems replete with microorganisms. In general, there are substantial differences between microorganisms found in the soil *versus* on animals. However, certain parasites and bacterial pathogens can survive at least temporarily in the soil and contaminate susceptible hosts upon contact. These potentially infectious agents typically display some form of latency or resistance. By slowing down or suspending their life functions, they can survive for long periods of time. For instance, this strategy is utilised by anthrax bacteria (*Bacillus anthracis*), which form spores; parasites such as taenids (tapeworms) and nematodes (roundworms), which produce eggs; and the pathogens responsible for toxoplasmosis and coccidiosis, which form cysts.

The risks associated with anthrax have long been recognised and conveyed via the ancient notion of "cursed fields". Cattle can end up dying from anthrax if they are put out to graze in pastures where the carcasses of previously infected cattle have been buried. We now know that earthworms, through their soil-churning actions, can move bacterial spores up to the surface over a period of many years. If grazing bovines consume the spores and spiny plants in tandem, they can develop an infection because microlesions caused by the plants allow the bacteria entry.

Bacteria in the genus *Yersinia* are also soilborne. Indeed, the best-known representative of this group — the bacterium responsible for plague (*Y. pestis*) — can survive for years in the burrows of rodents that have died of the disease. The environmental conditions underground are milder than those on the surface. If new rodents move into the burrow, they can become infected and launch a new disease cycle.

Tetanus is caused by the neurotoxin secreted by *Clostridium tetani*, a bacterium whose spores display great environmental resistance. Disease occurs when these spores contaminate a wound. Horses are particularly vulnerable to tetanus. Whether or not tetanus is a zoonosis is a topic of debate, given that the environment serves as the source of infection for both humans and other animals. However, animals are thought to act as reservoirs because the bacteria can be found in their digestive flora. Such is also the case for *Clostridium botulinum*, the bacterium responsible for botulism. Tetanus now rarely occurs in developed countries thanks to the availability of vaccines targeting the neurotoxin.

One specific opportunity for soilborne transmission in humans occurs when children play in sandboxes in public gardens or private residences. Indeed, domestic carnivores are attracted by these boxes, but for reasons other than play. Such areas become giant litter containers where animals defecate. Sandboxes can rapidly become wellsprings of bacterial and parasitic contamination for children if two conditions co-occur: a) there are many cats and dogs around, whether free roaming, feral, wild, or homeless and b) these play areas are not properly adapted, supervised, and maintained. Even if some of the parasites are fairly host specific, if a child swallows an infectious form, they can end up with an infestation (e.g., *larva migrans*). Sometimes the consequences are merely unpleasant. At other times, they are much more severe.

Foodborne Transmission

Foodborne zoonoses (or foodborne illnesses) result from the consumption of food contaminated with bacteria, bacterial toxins, viruses, and/or parasites. They are transmitted via the faecal-oral route: infectious agents are excreted by animals in forms that vary in their environmental resistance. They then occur in foodstuffs or water contaminated by the excreta. They reach the human body through the gastrointestinal tract. Most commonly, the initial symptoms provoked are vomiting and diarrhoea (sometimes the descriptor "stomach flu" is used). There are two general categories of foodborne illnesses. Foodborne infections occur when the illness results from the ingestion of food containing live bacteria or viruses that then establish themselves in the human intestinal tract. In contrast, foodborne intoxications occur when the illness results from the ingestion of toxins produced by bacteria during their growth in or on foods.

Foodborne illnesses can range from mild to extremely severe. Foodborne zoonoses are most often transmitted by fish, seafood, ham, salads containing eggs, and dairy products. They may also result from fruits and vegetables that have been poorly washed, washed with contaminated water, grown on land fertilised with animal waste, or irrigated with water contaminated by animal waste. Consequently, not even vegetarians are safe from food-borne zoonoses.

World travellers are quite familiar with the health phenomenon known as travellers' diarrhoea. It is most often caused by bacteria, sometimes by viruses, and occasionally by parasites. Local populations are frequently exposed to their local pathogens, generally resulting in natural immunity.

In Western countries, the risk of contracting foodborne diseases is smaller, thanks in part to the strict health and safety standards that are applied as foods travel from their point of production to their point of sale. It is also important to remember that food storage conditions within households can have a major impact. That said, smaller does not mean negligible. For example, in the European Union, there are approximately 315,000 confirmed cases of foodborne zoonoses each year. The actual number of cases is likely much higher. Indeed, government authorities are only aware of a small percentage of incidents: about 0.5% and 0.1% for bacterial and viral foodborne zoonoses, respectively. Thanks to reporting requirements for physicians, swift action can be taken to deal with outbreaks of foodborne illnesses, especially when food services are involved. Such outbreaks are defined as situations in which two or more people display similar symptoms, usually gastrointestinal, whose cause can be traced back to the same food or drink source. Overall, food safety alerts have a significant economic impact because they lead to hospital care for certain patients, launch action plans to identify the source of the outbreak, result in the recall of the affected lots of food, ensure the implementation of preventive or corrective measures, and promote communication with consumers. The most prominent example is perhaps the "mad cow" crisis (see p. 105). Foodborne zoonoses are thus clearly a major public health concern.

Food can be contaminated at any point along the path between producer and consumer. At present, health risks are amplified because of intensive livestock farming and the industrial trans-formation of animal products. Because a given facility may slaughter 10,000 chickens per hour or 10,000 pigs per day, we should not be surprised when pathogens spread like wildfire. Similarly, the risk of contamination is extremely elevated on food production lines that handle hundreds of different animal products. Furthermore, globalisation has given rise to public health incidents at the international scale.

Bacteria are most often responsible for foodborne zoonoses. The main species involved are *Campylobacter jejuni* and *C. coli*, which cause campylobacteriosis; *Salmonella typhimurium* and *S. enteritidis*, which cause salmonellosis (they are distinct from *S. typhi*, which causes typhoid fever); *Yersinia enterocolitica* and *Y. pseudotuberculosis*, which cause enteric yersiniosis (they are distinct from *Y. pestis*, which causes the plague); and certain strains of *Escherichia coli*. These bacteria naturally occur in the intestinal microbiota of wild animals and farm animals (e.g., poultry, ruminants, and pigs). However, they do not have any pathogenic effects in the latter. In contrast, they can cause very serious infections in humans, especially in vulnerable people. Such is often the case with strains of enterohemorrhagic *E. coli* (EHEC), which produce toxins that cause severe abdominal pain and bloody diarrhoea. The most problematic for public health is serotype O157:H7, which causes what is sometimes known as "hamburger disease".

Another example is *Listeria monocytogenes*, a bacterium found in the digestive tract of many vertebrate species, including bovines, sheep, goats, and chickens. It occurs in food contaminated by faeces from these species. This bacterium displays a rare trait: it can multiply at low temperatures and thus easily proliferates in household refrigerators. It can cause listeriosis, a rare but serious disease that shows up in vulnerable people.

Viruses are also quite often behind foodborne zoonoses. For example, *Calicivirus* and *Rotavirus* viruses can cause digestive disorders (gastroenteritis), while hepatitis E virus is responsible for the liver infection sometimes referred to as jaundice. Viral foodborne infections are primarily associated with seafood, fruits, vegetables, and undercooked meat (notably pork in the case of hepatitis E virus).

Finally, foodborne zoonoses can result from the consumption of poorly cleaned fruits and vegetables or animal products infested by the early developmental stages of parasites. One example is helminth eggs (e.g., from liver flukes, roundworms, or *Echinococcus* species), which display tremendous environmental resistance. Another example is provided by echinococcosis, a rare but serious zoonosis (see p. 85) that usually results from eating wild berries (e.g., strawberries or blueberries) or mushrooms contaminated by fox (*V. vulpes*) faeces. Similarly, protozoan parasites like *Giardia* and *Cryptosporidium*, found in the digestive tracts of humans and other animals, can contaminate food crops such as lettuce. The same is true for *Toxoplasma gondii*, which can cause toxoplasmosis in individuals who consume fruits or vegetables contaminated with cat faeces. Notwithstanding, the most common pathway for contracting toxoplasmosis is by eating raw or undercooked meat (see p. 87). Undercooked meat products can also cause macroparasite infestations in humans since the muscle tissue of animals may contain the larvae of parasitic worms such as taenids (more commonly known as tapeworms) and trichinae (see p. 88).

Fish can also be a problem when they harbour zoonotic parasites in the genus *Anisakis*. Sometimes called herring worm disease, anisakiasis is a zoonosis with a global distribution. Its infectious agent is present in the world's seas and oceans. It is well known in northern Europe and Japan, where populations frequently consume raw or marinated raw fish. Case numbers are currently low in France, but a few dozen of cases occur each year in Europe. Regulations require that raw fish be exposed to a freezing treatment (i.e., spend a minimum of 24 hours at a temperature lower than or equal to -20°C), which kills the parasite's larvae prior to consumption.

Waterborne Transmission

Hippocrates first identified water as a source of disease in the 4th century BCE. However, it was not until Pasteur's era that the first formal evidence of a causal relationship between water and diarrheal disease was established. It happened just a few years before Pasteur became interested in waterborne germs; he asserted, "We drink in 90% of our diseases". In 1854, English physician and epidemiology pioneer John Snow demonstrated that the cholera epidemic raging in certain parts of London traced back to a pumping station that drew its water from the Thames. When he convinced city authorities to render the pump inaccessible by removing the handle, the epidemic stopped, providing empirical evidence of the (non-zoonotic) disease's waterborne origin. Globally, access to drinking water remains a major concern. A 2012 UNICEF report stated that about 90% of deaths due to diarrheal disease worldwide are the result of poor water quality, inadequate sanitation, and inadequate hygiene practices.

Indeed, most of the zoonotic viruses, bacteria, and parasites involved in faecal-oral foodborne transmission (see p. 49) are resistant to external environmental conditions and can be passed along via contaminated drinking water. Not all diseases associated with drinking water are zoonoses: some are due to human-specific microbes (e.g., cholera, diphtheria, typhoid fever, hepatitis A, poliomyelitis, or norovirus gastroenteritis).

Some waterborne diseases are acquired after skin exposure. People frequently get cercarial dermatitis (i.e., swimmer's itch) after swimming in lakes and slow-moving waterways. This condition is caused by schistosome trematodes (genus *Tricho-bilharzia*) that parasitise waterfowl. They release microscopic larvae (cercariae, commonly referred to as "duck fleas") that sometimes penetrate the skin of swimmers. In the life cycle of these schistosomes, aquatic molluscs and birds serve as the intermediate and definitive hosts, respectively. Humans are only incidental hosts in whom the disease is benign; the cercariae stop migrating at the skin barrier upon entering an inappropriate host. However, humans act as definitive hosts for much more pathogenic schistosomes (genus *Schistosoma*) whose adult forms develop in the veins and can cause digestive or urinary disorders. As with cercarial dermatitis, the parasite's eggs are released into the water by the definitive host. A freshwater gastropod then serves as an intermediate host. The resulting zoonotic disease is called schistosomiasis, or bilharzia. Cases sometimes occur in Mediterranean countries. It is the second most common para-sitic disease in the world, surpassed only by malaria. In Europe, some unexpected outbreaks in 2014 were traced back to people who had gone swimming in a river in Corsica, France. Genetic analyses of the parasites traced their origins back to West Africa; they had arrived in schistosome-hosting visitors. New cases cropped up in 2018. It is not clear whether the source ensuring parasite persistence is humans or other animals. The challenge is to prevent the parasite from fully establishing itself in Corsica.

Leptospirosis is also a waterborne disease that people catch if their skin or mucous membranes are worn away or softened after extended periods in the water (see p. 68).

Finally, water serves as a breeding ground for the larval forms of many vectors, especially mosquitoes. Consequently, effectively managing wetlands is essential in the fight against zoonoses.

Airborne Transmission

Erik Satie[1] once sardonically commented, "Don't breathe without first boiling the air", referring to the recommended practice for raw milk. Indeed, many zoonotic diseases are transmitted via the air. Transmission can be direct when people inhale solid or liquid airborne particles containing microorganisms. It can also be indirect, if one person's respiratory mucosa come in contact with someone else's infectious respiratory secretions, which can occur because of inadequate hand hygiene or the handling of soiled objects. People's hands or the soiled objects serve as fomites, materials capable of passively transmitting infection. Public health campaigns have greatly focused on these two modes of transmission in efforts to limit the spread of SARS-CoV-2.

Airborne particles are classified according to size, and their size determines how deep they can penetrate into the body (see Figure 6). Particles larger than 10 µm in diameter remain in the upper airways (nose, mouth), while particles smaller than 10 µm in diameter (inhalable particles) can enter the bronchi. Fine inhalable particles (< 2.5 µm) can penetrate deep into the respiratory tract and reach the pulmonary alveoli. Finally, very fine particles (< 1 µm) and ultrafine particles (< 0.1 µm; also called "nanoparticles") can pass through the alveolar-capillary barrier and enter the bloodstream.

---

1. In *Écrits*, collected by Ornella Volta, Éditions Champ Libre, 1981.

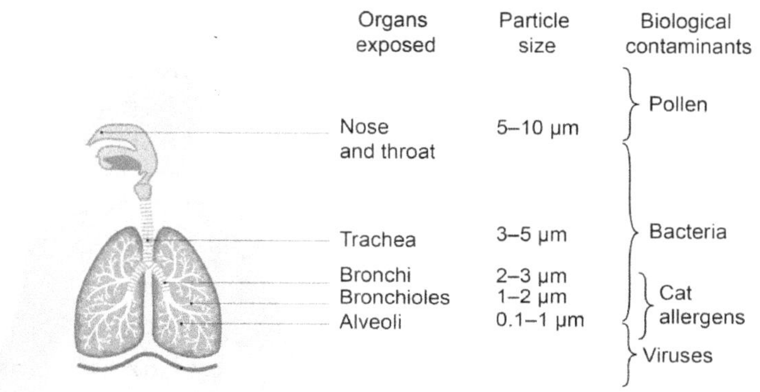

Figure 6. Depth that biological contaminants penetrate into the respiratory tract as a function of size.

Some pathogens can overcome or even exploit the defences of our respiratory system. An example is provided by the zoonotic bacterium *Coxiella burnetii*, which causes Q fever. It infects the macrophages responsible for swallowing errant particles in the deep respiratory system. Not only can *C. burnetii* withstand the acidic pH within the macrophages' vacuoles, but it can also use the macrophages to multiply and enter other parts of the body. This bacterium is excreted by female ruminants, mainly during the birthing process, and it consequently occurs in manure. When the latter is spread as fertiliser, airborne transmission to humans may result (see p. 74). Another intracellular bacterium, *Chlamydia psittaci*, is asymptomatically found in birds. When it is transmitted to humans, it can cause psittacosis, a potentially serious respiratory disease (see p. 71).

Viruses are able to exploit molecules (e.g., receptors) that are expressed on respiratory surfaces, which allows them to bind to their target cells. When viruses display pathogenicity in tandem with efficient airborne transmission among humans, they have the potential to provoke a pandemic. The worldwide spread of pandemic influenza viruses (see p. 96) and COVID-19 shows that such risks are quite real.

Airborne pathogens also make frightening candidates for biological weapons, given that they can affect large numbers of people quite quickly. They could be spread via spraying, diffused by air conditioners, or dispersed using an explosion. In particular, certain zoonoses could be effectively exploited by bioterrorists, including anthrax (see sidebar p. 45), tularaemia (see p. 44), Q fever, typhus (caused by bacteria in the family Rickettsiaceae), glanders (*Burkholderia mallei*), and plague (*Y. pestis*).

Vector-borne transmission

Mosquitoes, fleas, and ticks are small animals that are a large nuisance because their bites are itchy and may provoke painful reactions or allergies. However, their key importance in terms of public health and economics stems from their ability to transmit pathogens. They are thus known as vectors, or "organisms that can transmit infectious pathogens between humans, or from animals to humans", as defined by the WHO.

Most vectors are arthropods, either insects, such as mosquitoes, phlebotomine sand flies, black flies, biting midges, and fleas, or acarians, such as ticks. Over the course of their lives, these arthropods take several blood meals. When feeding on an infected host, they may end up ingesting an infectious agent along with the blood. This pathogen will then be passed on to another host when the arthropod takes another meal. Zoonotic transmission occurs via vectors that feed on both humans and other animals. When a pathogen is transmitted by a female vector to her eggs, the vector is also acting as a reservoir. It can thus maintain the transmission cycle on its own (see p. 14).

The effectiveness of pathogen transmission by vectors depends on multiple factors: the vector's host preferences; how well the pathogen is adapted to different hosts; and environmental conditions, which affect survival and mediate encounters between vectors, humans, and other animals. For some pathogens, such as West Nile virus or the bacterium responsible for Lyme disease, the transmission process must involve animals other than humans. The latter are epidemiological "dead ends". In other words, even if humans become ill, they do not pass the pathogen along to a new vector. For others, such as the chikungunya, dengue, and yellow fever viruses, the transmission cycle requires non-human animals in some environments, notably tropical forests, but not in others, such as urban areas.

The world's number 1 vectors are mosquitoes (family Culicidae). There are more than 3,500 species in the world. Only some bite humans. Bites are generally caused by female mosquitoes — they require a blood meal prior to laying for egg development to proceed properly. The most notorious disease in the world is malaria: more than 200 million people across the globe are infected. It is caused by parasites in the genus *Plasmodium* that are vectored by mosquitoes in the genus *Anopheles*. However, malaria is not zoonotic, except in the case of two species (see p. 17). By observing *Plasmodium* parasites in mosquitoes and comparing the geographical distribution of *Anopheles* with that of malaria, scientists were able to establish the role of mosquitoes as vectors in the late 19th century. Mosquitoes are also involved in the transmission of several viruses belonging to various families; the umbrella term arbovirus is used to refer to this group of pathogens (an acronym for arthropod-borne virus). Of particular note are West Nile virus, which occurs sporadically in the Mediterranean Basin, and the yellow fever, dengue, chikungunya, and Zika viruses, which are naturally found mainly in tropical and subtropical areas. However, local transmission of some of these viruses is beginning to occur in temperate zones, due to the expanded range of one of their vectors, the tiger mosquito (*Aedes albopictus*). In France, it is possible for the general public to report observations of the tiger mosquito via an online portal; these data help build knowledge about the species' distribution.

Ticks are also major vectors. They feed on their hosts for extended periods of time and transmit a wide range of potential pathogens: bacteria, viruses, and parasites. Their development is much slower than that of mosquitoes, with life cycles that span several years. A distinction is made between "hard" ticks (~700 species worldwide), which have a hard dorsal plate on their exoskeletons, and "soft" ticks (~180 species), which have soft exoskeletons. Soft ticks can transmit bacteria in the genus *Borrelia*, which cause recurrent fevers, especially in Africa. Hard ticks, whose bites are painless, are a much larger health concern than are soft ticks. In western Europe, the main tick-vectored zoonosis is Lyme disease. Ticks also transmit viral encephalitis, such as tick-borne encephalitis, which is endemic in different parts of Europe. Additionally, they pass along the agents responsible for viral haemorrhagic fevers, such as Crimean-Congo haemorrhagic fever. Several cases of the latter have been reported in Spain and Bulgaria. Multiple European countries have developed digital apps that people can use to report tick bites (e.g., *Signalement TIQUE* in France); the data will promote understanding of the risks associated with these vectors.

Although fleas are adapted to specific hosts, they may sometimes feed on other animals, which can result in the transmission of zoonotic agents. The best-known example is plague, which does not currently occur in Europe. However, it can be found in tropical countries, and it causes local clusters of infections in the western US. Fleas also vector *Bartonella henselae* in cat populations, a bacterium that can be transmitted to humans, most often via cat scratches.

There are other species capable of vectoring zoonotic agents, which are mainly found in tropical or subtropical areas. Phlebotomine sand flies are small insects that look like midges. They are widely distributed around the Mediterranean as well as across much of Africa and the Middle East. They transmit the infectious agent responsible for leishmaniasis (see p. 62). In Africa, tsetse flies transmit *Trypanosoma brucei*, which causes sleeping sickness. Cattle are one of this pathogen's reservoirs. In the Americas, a different trypanosome, *T. cruzi*, causes Chagas disease and is transmitted via the infected droppings of kissing bugs (i.e., triatomines). More than 180 mammal species can serve as its reservoir, especially dogs, cats, rats, and a variety of wild animals.

The distributions of vector-borne diseases are shifting due to climate change, habitat modification, biodiversity loss, international trade, and the movement patterns of animals and people. As a result, diseases such as dengue fever, West Nile fever, and chikungunya have appeared in countries where they were not present before. International trade has facilitated the introduction of these pathogens into new areas. Climate change is capable of promoting their persistence and shifts in their seasonality.

MAJOR ZOONOSES BY HOST TAXON

In this section, we discuss examples of zoonoses transmitted to humans by our closest animal relatives (i.e., primates); animals that are behaviourally tied to us (i.e., domestic and commensal animals); and animals with which we are in increasing contact due to habitat destruction (e.g., bats).

Non-human Primates

As of 2018, there were 518 known species of primates. Non-human primates are mainly found in tropical regions. They are mostly arboreal and frugivorous, with some exceptions, and deforestation is driving many extinct. Almost all have been granted protected status in the countries where they occur. No non-human primates are native to Europe; the Barbary macaque (*Macaca sylvanus*) occurs in the wild in Gibraltar but is an introduced species. Thus, any contacts with these animals result from interactions in research and experimental laboratories, zoos, zoo-like establishments, or private homes, where they are often illegally kept. While EU legislation requires that health and safety standards be met prior to the importation of any non-human primates, existing requirements remain inadequate given known health concerns.

Primates display an astounding degree of diversity, ranging from the tiny Malagasy mouse lemurs (*Microcebus* species) and South American marmosets (*Callithrix* species) to the enormous African gorillas (*Gorilla* species). It is important to note that monkeys are not the only primates. The order Primates is divided into two suborders: Strepsirhini, which contains the lemurs and lorises, and Haplorhini, which contains the tarsiers and monkeys. The monkeys (infraorder Simiiformes) are themselves separated into two major groups: the New World simians *versus* the Old World simians. The two can be easily distinguished by nostril spacing, which is greater in the former than the latter.

Humans are Old World simians. We are members of the family Hominidae, as are the gorillas, chimpanzees (*Pan* species), and orangutans (*Pongo* species). The two other families in our infraorder are Cercopithecidae, composed of the macaques (*Macaca* species), the langurs (*Semnopithecus* species), and the cercopithecids (*Cercopithecus* species), and Hylobatidae, composed of the gibbons.

Zoonotic risks vary depending on species identity and location. The latter may be the natural environment, research laboratories, zoos, or private households. The origins and health statuses of non-human primates found in research laboratories tend to be well known because the animals are essentially bred in captivity. Indeed, only a small proportion are members of introduced species (mostly Asian macaques from breeding farms). However, the latter represent the most commonly imported laboratory animals: for example, around 2,000 or 3,000 individuals are brought into France each year. In the European Union, cats, dogs, and non-human primates represent 0.25% of the 10.6 million animals used in research, whereas mice, fish, rats, and birds represent 92%. Establishments that publicly display animals import very few annually, but the species they bring in can be quite varied. The health status of those animals is subject to rigorous standards. When it comes to animals purchased by private owners, little is known, and there are few to no guarantees about the animals' health.

Viral diseases are probably the greatest cause for concern. Among the most emblematic examples is rabies. In the late 1980s, the illegal importation of Barbary macaques (*Macaca sylvanus*) from Algeria to France resulted in several cases of vaccination-induced rabies. Before leaving Algeria, the macaques were given a live attenuated vaccine that is inappropriate for the species and forbidden in France. Fortunately, no human cases resulted. Another virus of great concern is herpesvirus simiae, or monkey B virus. Even if a very small number of human cases have been observed worldwide (~a few dozen), the case fatality rate is high (80%). The virus is found in all Asian macaques — all the *Macaca* species — with the exception of the North African macaque. It is worth noting that the *M. fascicularis* population on Mauritius, introduced in the 16th or 17th century from Southeast Asia, is also free of the disease. In contrast, seropositivity ranges from 70 to 100% in wild populations in Asia. Primates also play a key role in the sylvatic transmission cycles of arboviruses, such as those that cause chikungunya, dengue, and Zika. Additionally, concern is growing over the agents responsible for viral haemorrhagic fevers, also found in non-human primates. Depending on the situation, they are either reservoirs (Marburg virus disease and yellow fever) or victims (Ebola in Africa and Ebola-Reston in Asia) (see p. 101).

With across-the-board declines in non-human primate species, we are discovering the role played by these animals in the epidemiology of potentially dangerous viruses, bacteria, and parasites. However, for many diseases (e.g., Ebola or mycobacterioses), non-human primate species are the hardest hit by epizootics.

Cats and Dogs

The order Carnivora contains around 300 species, including both cats and dogs, members of the families Felidae and Canidae, respectively. Cats have shared our lives for at least 9,000 years. For dogs, this figure is closer to 15,000 years. They can transmit diseases to us in multifarious ways but most commonly via faecal contamination, contact with their fur, bites, scratches, or shared arthropod vectors. Transmission can also result under highly specific conditions, such as when an animal gives birth or sports a wound. That said, it is generally possible to limit disease transmission risks by practicing rigorous basic hygiene and maintaining your pet in good health.

Human infections mainly arise from contact with animal excrement, which can contain helminth ("worm") eggs, protozoa, and bacteria. Among the helminths are roundworms, such as ascarid nematodes and hookworms, and flatworms, such as taenids and *Echinococcus* tapeworms, that are naturally hosted in the intestines of cats and dogs. When ingested by humans, the parasites' eggs or larvae can migrate into various organs and cause symptoms of varying severity (see p. 49). Children are at greater at risk of contracting these parasites because they frequently play on the floor, "put everything in their mouths" (see p. 47), and cuddle with animals, an intimate degree of contact. It is essential to regularly deworm cats and dogs to limit transmission within households. Cats also naturally host the protozoan responsible for toxoplasmosis, which especially presents risks during pregnancy.

Generally, contact with the skin or coat of cats and dogs is not particularly problematic, as long as the animals are healthy and regularly treated with external antiparasitics. The only concern is that the animals' fur could be contaminated with faeces, which can contain transmissible pathogens. It is important to keep an eye on animals that are scratching themselves a lot or losing their fur: they may have fleas, ringworm, or scabies (see p. 44). Similarly, an animal with a runny nose or watery eyes may be hosting bacteria that can cause conjunctivitis in humans. In any of these situations, you can see a veterinarian to establish a diagnosis and determine whether there is a risk of transmission to humans. You should remain very vigilant when bites and scratches occur, given the health issues that can arise (see p. 47).

Cats and dogs may be bitten by arthropod vectors and can act as reservoirs for a number of pathogens. For example, phlebotomine sand flies are found in the Mediterranean and can transmit *Leishmania infantum*, a protozoan that may provoke leishmaniasis in dogs. This parasite mainly causes problems in children and the immunocompromised. Serious cutaneous or digestive conditions may result, which can be fatal if left untreated. Normally, dogs do not transmit ticks directly to humans: both dogs and humans are targeted by ticks found outdoors on vegetation, where they lie in wait for potential hosts.

Bovines, Sheep, and Goats

Bovines, sheep, and goats are members of subfamilies within the family Bovidae, which, in turn, is part of the order Cetartiodactyla. There are around 550 cetartiodactylan species, whose taxonomic distinction is possessing an even number of digits. Among the bovines are several major livestock species, notably taurine cattle (*Bos taurus*), zebu cattle (*Bos indicus*), yaks (*Bos grunniens*), and buffalo (*Bubalus* spp.). Within the sheep are both domesticated species (*Ovis aries*) and wild species (also in genus *Ovis*). Similarly, there are domesticated goats (*Capra hircus*) and wild goats, such as the Alpine ibex (*Capra ibex*). Current species of bovine livestock are descended from the now extinct aurochs (*B. primigenius*), which was domesticated at least 10,500 years ago at two distinct locations (seep. 39). Goats were domesticated around 10,000 BCE in Iran, and sheep were domesticated around 8,500 BCE in the Middle East and India. Currently, there are around 1.5 billion head of bovine livestock in the world.

Bovines, sheep, and goats are ruminants, a term that refers to their unique digestive system. It comprises several compartments that are structured so as to promote digestion via rumination. During this process, animals rechew previously ingested food. More specifically, the food is regurgitated, newly mixed with saliva, masticated a second time, and then reingested. Domestic ruminants provide humans with vast quantities of meat, dairy products, leather, and fertiliser (i.e., manure). They also provide environmental services: they may maintain grasslands, clear natural spaces, and provide manual labour.

With regards to zoonoses transmitted by bovines, the largest disease prevention programmes have successfully targeted major diseases of historical importance, such as bovine tuberculosis (infectious agent: *Mycobacterium bovis*) and brucellosis (infectious agents: *Brucella* bacteria), even if their complete eradication remains complicated. In 2020, 86 cases of bovine tuberculosis were reported in the European Union. Examples of foodborne zoonoses associated with beef consumption are EHEC (see p. 51) and bovine spongiform encephalopathy. The latter disease is now well under control (see p. 105).

The people most exposed to the zoonoses transmitted by domestic ruminants are those who interact with them on a daily basis, such as livestock farmers, veterinarians, and slaughterhouse workers. Pathogens can be passed along when humans come in contact with an animal's skin, carcass, or mucous membranes or when they inhale contaminated dust or air. Events like births and abortions particularly entail health risks because the placenta can transmit various zoonotic bacteria, such as those that cause Q fever. Consequently, when farms are open to the public, it is recommended that birthing females be placed in specific rooms where visitors are not allowed or that visits be limited during the birthing period.

Pigs and Wild Boars

The pig (*Sus domesticus*) was domesticated from the wild boar (*Sus scrofa*). Both belong to the family Suidae, which is part of the order Cetartiodactyla, like the family Bovidae. The pig was domesticated independently in at least two different locations: in northern Mesopotamia (present-day Iraq) around 7,500 BCE and in China around 6,000 BCE. The pig and the wild boar are extremely close relatives, so close that they can reproduce with each other. Because of this proximity, they jointly contribute to various health risks. In pigs, changes in livestock farming practices over time have certainly altered the animal's impact on public health in major ways. There are clear differences in the hazards posed by traditional farms, which sometimes still have outdoor pens in Europe, and those posed by large, exclusively indoors industrial farms. We must not forget the potential issues arising from pet pigs, which represent a growing segment of the pet population.

Pigs have been associated with various zoonotic viruses, bacteria, and parasites. Hepatitis E virus (family Hepeviridae, genus *Orthohepevirus*) is zoonotic but has a life cycle that comprises several possible pathways. Infection can occur via pork consumption, but also via direct contact with live animals or contaminated water. Furthermore, pigs are not the virus' only reservoir.

Influenza virus species also readily circulate among pigs. It is known that the risks of new viruses emerging climbs when humans, pigs, and domestic ducks come together. Epidemiologists hypothesise that viral recombination and, consequently, zoonotic virus emergence is more likely in regions of the world (e.g., Asia) where wild ducks crossbreed with domestic ducks reared outdoors that also come into contact with pigs.

While swine brucellosis (infectious agent: *Brucella suis*) has practically disappeared from industrial pig farms, it is still found in wild boar populations. Indeed, there were unseen health consequences that arose from the push for free-range pig farming in the late 20th century, an effort intended to promote animal welfare. While the farms were fenced, preventing sows from leaving, this system did not always prevent male boars from visiting. Livestock farmers ended up with litters of boar-pig hybrids and animals infected with swine brucellosis. Although human cases have resulted from direct contact with wild boar carcasses, they remain rare. Interestingly, specific *B. suis* strains have been found to circulate independently among European hares (*Lepus europaeus*).

Pigs and wild boars can harbour two parasites: a tapeworm (class Cestoda, *Taenia solium*) and a roundworm (class Nematoda, genus *Trichinella*; see p. 88). *T. solium* is a highly emblematic tapeworm species. It occurs as a mature, reproductive adult in the intestines of humans and as infectious larvae (i.e., cysticerci) in pigs. Its presence has significant impacts on host health. Control programmes at the community, regional, and national levels have shown that the parasite's transmission cycle can be interrupted by carrying out tandem treatments in humans and pigs, while also respecting proper hygiene practices. This approach has yielded promising results in countries as different as Peru and Zambia.

Horses

Horses (*Equus caballus*) are part of the family Equidae in the order Perissodactyla. This order comprises 21 species whose taxonomic distinction is having an odd number of digits on their hind limbs. There are three families. Equidae contains horses, donkeys (*Equus asinus*) and related species, and zebras (*Equus grevyi*) and related species, which all have a single visible digit on their four limbs. Tapiridae contains species with three digits on their hind limbs and four digits on their forelimbs. Finally, Rhinocerotidae contains species with three digits on their four limbs. Humans domesticated horses after cattle, around 4,500 BCE on the Eurasian steppes.

The main zoonoses passed from equids to humans arrive via skin contact with animals infected with ringworm or scabies (see p. 44) or as a result of poor hygiene when handling dung. The most serious diseases are currently infrequent in Europe thanks to the implementation of control measures. Examples include pseudotuberculosis (*Yersinia pseudotuberculosis*), anthrax (*B. anthracis*), and glanders (*Burkholderia mallei*), a disease that has almost been eliminated from North America, Australia, and most of Europe. The latter two bacteria are classified as potential bioterrorism agents because they are infectious at low doses and can be spread via aerosols. It is also important to note that horses played a role in the emergence of Hendra virus in Australia (see p. 68).

Additionally, horses can be infected by the arboviruses responsible for viral encephalomyelitis, including West Nile virus in the Mediterranean Basin, Japanese encephalitis virus in Asia and Oceania, and the American equine encephalomyelitis viruses in the Americas. Depending on the virus, the reservoir may be a bird or rodent species. The horse is an epidemiological dead end (i.e., it does not retransmit the virus to the mosquito vector). Horses are therefore not a source of infection, but they may act as sentinels and signal when a virus is circulating locally. Finally, horses may convey the relative risk of tetanus in an area, given that they are extremely sensitive to the disease.

Rodents

By far, most mammal species are found in the order Rodentia (rodents), followed by the order Chiroptera (bats). Of the 6,495 known mammal species in 2018, rodents accounted for 2,552 (39%). Although the exact numbers and percentages vary somewhat depending on the source, the sheer abundance of rodent species may result in a greater diversity of pathogens than are found in other mammalian orders. Although this list is not nearly exhaustive, we discuss some key zoonotic pathogens of major importance to public health.

The viral zoonoses with the most notable impacts on humans are caused by agents in the families Hantaviridae, Poxviridae (monkeypox and cowpox, genus *Orthopoxvirus*), and Arenaviridae.

Hantaviruses cause haemorrhagic fever with renal syndrome in the Old World and respiratory diseases in the New World. The latter were only discovered in the 1990s. Across the world, many rodents host species-specific hantavirus strains, and a certain amount of coevolution has occurred between rodents and the viruses they carry. Furthermore, each pair displays a fairly specific geographical distribution. That said, Seoul virus and its host, the Norway rat (*Rattus norvegicus*), have gone cosmopolitan. The virus is found across the globe, especially in large cities where the rat is well established. Public health threats generally arise in one of two ways. First, humans may be exposed to rodents in rural areas or zones undergoing deforestation. Second, localised infestations may result in rodents invading human-occupied spaces (e.g., houses and/or villages), where the animals temporarily adopt a commensal lifestyle. In Europe, Puumala virus is hosted by the bank vole (*Clethrionomys glareolus*). While the pathogen's prevalence is highest in northern Europe (e.g., Finland), enzootic foci also exist, such as in the Ardennes mountains along the border between France and Belgium.

Poxviruses occur in many vertebrate species and are particularly well characterised in mammals. The only example of a human-specific poxvirus is smallpox, which was eradicated via vaccination campaigns in the late 20th century. There are many other viruses in the same family, including many that are hosted by rodents even if their names suggest otherwise. Cowpox virus occurs in field rodents in Europe, and monkeypox virus is found in various rodents in Africa. The circulation of monkeypox is currently being monitored by the WHO. Indeed, given that younger generations are no longer vaccinated against smallpox, epidemiologists have speculated that the virus' vacant ecological niche could be filled by another poxvirus, notably monkeypox.

The arenaviruses are probably the least well known of this triad. Examples of arenaviruses are the agents responsible for Lassa fever and lymphocytic choriomeningitis. The latter appears to be passed to humans by pet hamsters and mice. When the infection occurs during pregnancy, it can have serious effects on the foetus. Lassa virus is currently circulating in West Africa, particularly in Nigeria, where it has caused a marked number of infections and deaths over recent years. The virus is hosted by murids in the genus *Mastomys* (multimammate rodents), which are not usually synanthropic. However, depending on where they are in their population cycle, they may move closer to human homes and granaries, increasing zoonotic risks.

Among the bacteria transmitted by rodents, three classic examples are *Y. pestis* (plague), *Leptospira* species (leptospirosis), and *Brucella suis* (brucellosis). Leptospires occur in the urinary tracts of many rodent species, which shed the bacteria in their urine. Amphibious species, such as Norway rats, muskrats (*Ondatra zibethicus*), and nutria (*Myocastor coypus*), can contaminate the water bodies, rivers, canals, and wetlands they inhabit. All three species have been introduced to Europe. Leptospires cannot penetrate healthy skin, but they can pass through abraded skin or skin that has been softened after soaking in the water. In France's tropical overseas departments, infection risks may be high in certain croplands, such as those covered by rice or sugar cane, because agricultural workers are sometimes improperly equipped. In mainland France, leptospirosis is officially recognised as an occupational disease for sewage workers. The annual incidence of leptospirosis has climbed since 2014 without any clearly identifiable cause. These patterns of occupational risk hold at the European level. Thus far, the disease has mainly been seen in Mediterranean and Eastern European countries.

Bats

To date, 1,386 known bat species have been described (21% of all mammals), making Chiroptera the second most species-rich mammalian order after Rodentia. In 2019, the International Union for Conservation of Nature (IUCN) estimated that 15% of bat species were facing extinction. As in the case of rodents, the sheer number of bat species partly explains why they harbour a wide range of pathogens. Bats are present on all continents, with the exception of Antarctica. Most species utilise echolocation, a sonar system, to navigate through the environment: they send out calls and exploit the returning echoes to determine the positioning of land features.

The bat species found in Europe are very small (5–45 g) and exclusively insectivorous. Asia and Oceania are also home to flying foxes (i.e., *Pteropus* species), which are frugivorous and among the largest bats in the world. Flying foxes can weigh as much as 1 kg, have wing spans of more than 1.20 m, and do not use echolocation. Other species within the same family (Pteropodidae) can be found in Africa. However, they are absent from the Americas. Bats can migrate several thousand kilometres. The Neotropics are the only region of the world to host haematophagous species (vampire bats), of which there are three.

Like all wild animals, bats carry many species of enteropathogenic bacteria (see p. 49). They are also infected by microscopic fungi (notably *Histoplasma* species), which are shed in their guano. These fungi can infect humans via airborne transmission, such as during cave visits, and can cause respiratory problems in unprotected or immunocompromised individuals.

Bats are also reservoirs for several rabies viruses (*Lyssavirus* species), which are not the same viruses that cause rabies in flightless mammals. However, the effect on humans is the same: the disease is fatal if not treated quickly. Infection can result when humans are bitten, scratched, or licked by bats. Bats with rabies may display altered behaviour, such as struggling to fly or letting humans approach them. Therefore, you must never touch an injured, dead, or strangely behaving bat. If you see such an animal, contact a specialist, such as a bat biologist or a veterinarian, who can professionally assess the situation.

The virus responsible for COVID-19 (SARS-CoV-2) is among the emerging viruses for which bats are potential or known reservoirs. Another example is Hendra virus. It was described for the first time in 1994 in Australia, when a number of horses and their caretaker suddenly died from an acute respiratory distress syndrome. The horses became infected after consuming fruit contaminated with the urine and saliva of *Pteropus* species. The horses then passed the infection on to their caretaker. Flying foxes can also transmit Nipah virus, which has a case fatality rate of 40–75% in humans. The first Nipah virus outbreak was observed in the late 1990s in Malaysia, where bat habitat was being destroyed as land was deforested to build palm oil plantations. Bats ended up foraging for food on pig farms where there were also fruit trees. They transmitted Nipah virus to the pigs, which subsequently infected humans. Since then, Nipah epidemics have occurred regularly in India and Bangladesh. In some cases, humans have been directly infected after consuming contaminated palm syrup or fruit. Similarly, fruit bats are hypothesised to be the wild reservoirs of filoviruses, responsible for cases of Marburg virus disease and Ebola in Africa, although data are still being gathered (see p. 101).

Finally, insectivorous bats (*Rhinolophus* or *Taphozous* species) are thought to have played a role in the emergence of the SARS-CoV-1, SARS-CoV-2, and MERS coronaviruses (see p. 89).

ARE BATS SUPER RESERVOIRS?

Nipah virus, Hendra virus, Ebola virus, SARS-Cov-1, SARS-CoV-2… whenever a viral pathogen emerges, bats always seem to be involved. However, when we look at the proportions of zoonotic viruses hosted by bats *versus* rodents, for example, there are no striking differences. So, is it really fair to say that bats are "super reservoirs" for viruses? Numerous studies have explored this question and identified unique characteristics in bats that could promote virus emergence. Bats are long lived for mammals (life expectancy: 20–30 years), which means they might potentially encounter more pathogens. Moreover, some live in huge colonies, containing thousands or even millions of individuals, which facilitates the spread of viruses. Most intriguingly, bats are the only mammals that are active flyers. Flight is metabolically costly and may thus be associated with the development of highly efficient mechanisms for repairing DNA and mitigating oxidative stress. Such mechanisms could have heightened their virus tolerance, increased their longevity, and afforded them protection against cancer.

Regardless, it is important to underscore that the emergence of viral diseases is not caused by bats' specific characteristics but rather by the environmental upheavals that destroy their habitats and modify the interfaces at which they come into contact with humans. Protecting bat species is essential because they play crucial roles in ecosystem functioning, such as regulating insect populations, pollinating plants, and dispersing seeds.

Birds

Class Aves is even more species rich than class Mammalia: to date, around 11,500 bird species have been described. Half are members of the order Passeriformes (i.e., perching birds/songbirds). We are still far from having characterised all the pathogens they may harbour.

Wild birds and humans share the unique ability of being able to rapidly travel long distances. During the annual migration period, billions of birds change continents to reach either their wintering grounds or nesting sites, depending on the season. Migrating birds carry along with them a whole range of viruses, bacteria, and parasites. Some are potentially pathogenic in humans, causing three classes of symptoms: digestive, respiratory/mucocutaneous, and nervous. Birds likely contribute to the broad distributions of many pathogens. On shorter timescales, bird migration patterns may lead to the occasional introduction of pathogens into entirely new areas and result in the emergence of new diseases in local populations.

Bird droppings contain pathogens that cause gastroenteritis, such as bacteria in the genera *Salmonella*, *Campylobacter*, *Yersinia*, and *Escherichia coli* and as well as parasitic *Giardia* species. Certain birds occupy the same habitats as humans. Notably, gull species frequently forage at landfills or sewage treatment plants and are thus exposed to the antibiotic-resistant bacteria found in waste. They can contribute to the spread of such bacteria within the environment. Bird droppings may also contain an intracellular bacterium, *Chlamydia psittaci*, which causes psittacosis in humans. This disease provokes flu-like symptoms that can result in severe lung infections. Birds, especially web-footed species, host the bacterium without being afflicted with the disease. Consequently, people working at indoor poultry rearing operations and duck slaughterhouses are particularly at risk of exposure.

In 2021, the FAO estimated that the global chicken population (*Gallus gallus*) was made up of over 30 billion individuals. It is estimated that there are 50 billion wild birds in the world. Soon, the number of poultry will equal the number of wild birds. *Influenzaviruses* are the most feared viruses found in domestic birds. They have major impacts on public health because they are responsible for the flu. However, contrary to what is sometimes reported in the media, it is extremely rare for influenza viruses to move directly from wild birds to humans. Although birds frequently host influenza viruses and shed them in their droppings, avian flu viruses cannot be transmitted to humans. In the very rare instances that they are, they may only cause conjunctivitis. The highly pathogenic influenza virus that appeared in domestic poultry in 2003 — the H5N1 virus — is an exception. It is highly pathogenic to both humans and wild birds (see p. 96).

Wild birds are reservoirs for zoonotic viruses and bacteria transmitted by arthropod vectors (see p. 56). For example, birds host the bacteria that cause Lyme disease (i.e., vectored by ticks) and the various arboviruses that cause encephalitis in humans and equids (i.e., vectored by mosquitoes), including West Nile virus, St. Louis encephalitis virus, Japanese encephalitis virus, and the Eastern and Western equine encephalitis viruses. Of these viruses, only West Nile virus is present in Europe. Also present is Usutu virus, which was first observed in Africa. It mainly affects songbirds, in which it has caused several mass mortality events in Europe since the mid-2010s. Blackbirds (*Turdus merula*) have been particularly impacted. However, Usutu virus is rarely and sporadically passed along to humans.

Birds harbour microscopic fungi (*Cryptococcus*, *Histoplasma*, *Candida*, *Aspergillus* species) in their plumage, beaks, and digestive tracts. These fungal species can cause respiratory or skin infections in humans, especially in young children, the elderly, and the immunocompromised. The latter group notably contains people with HIV or individuals undergoing chemotherapy for cancer.

Finally, despite their name, *Mycobacterium avium* complex (MAC) bacteria are not exclusively found in birds. They are actually microbes that live in the environment but that can infect birds, which then excrete them. MAC bacteria are mostly associated with traditional and family-run livestock farms. In humans, exposure can cause non-tuberculous disease in vulnerable individuals.

SOME EXAMPLES OF ZOONOSES

In the previous chapters, we defined zoonoses and described the circumstances that favour the transmission of viruses, bacteria, and parasites from animals to humans. Here, we explore 15 zoonoses, which illustrate certain important overarching issues. Plague, tuberculosis, and rabies illustrate the historical and modern-day importance of zoonoses. Tuberculosis in particular underscores that humans and other animals can swap pathogens back and forth. Lyme disease, coronavirus-caused illnesses, Crimean-Congo haemorrhagic fever, influenza, Ebola, and "mad cow" disease (a prion zoonosis) — all made headlines in the late 20th and early 21st centuries. They highlight the risks of "new" diseases arising. Yellow fever and West Nile fever show how diseases can spread worldwide thanks to the forces of international trade. Q fever and toxoplasmosis serve as reminders that health risks are amplified during vulnerable periods of life, such as pregnancy. Echinococcosis and trichinellosis underscore dietary risks and show that parasite cycles can be highly diverse. The emergence of antibiotic resistance is another example of the inextricable links between animal and human health.

BACTERIAL ZOONOSES AND ANTIMICROBIAL RESISTANCE

Q Fever

Q fever was first described in the 1930s in Australia, when a febrile illness of unknown origin spread through slaughterhouse workers. The disease was thus named "Query" or Q fever because its origin remained under investigation. It is now found everywhere in the world with the exception of New Zealand.

At the time, it was complicated to identify the infectious agent, the bacterium *Coxiella burnetii*, because it only grows in cells and, therefore, could not be cultured using classical methods. *Coxiella burnetii* can form resistant pseudospores that display high levels of environmental persistence. Infections mainly occur via airborne transmission, and some countries have classified it as a potential biological weapon. The bacterium can travel long distances when in aerosol form, as illustrated by the 1996 epidemic in the city of Chamonix, France, where the pathogen's spread seems to have been boosted by the presence of a heliport near a ruminant slaughterhouse.

In animals, Q fever primarily occurs in domestic ruminants (e.g., bovines, sheep, and goats), which are thought to be the bacterium's main reservoir. In these species, the disease largely manifests itself via spontaneous abortions. Infected animals excrete the bacteria in large quantities, especially in their faeces and in birth and abortion products.

If humans come in contact with these materials, they may become infected in turn. Often, there are no symptoms of disease, and the bacterium's presence is only discovered after the fact via targeted antibody testing. When symptoms are present, they can range from a simple flu-like syndrome to severe disease, including heart, lung, and liver problems as well as miscarriages. Clusters of infections sporadically emerge, where people are all infected via the same animal source. Q fever became a disease of concern in Europe when an exceptionally large outbreak occurred in the Netherlands in 2007–2010, during which 4,026 cases were reported. In 2020, there were 523 confirmed cases in humans in Europe.

Prevention requires monitoring the health of ruminant herds and watching for waves of abortions. Farmers should use the same precautions as with other abortive diseases — only handle aborted animals and placentas with gloves and dispose of these tissues at a rendering plant, actions that are often difficult to carry out. Furthermore, manure should be composted prior to spreading to reduce pseudospore viability. Finally, farms that are open to the public should prohibit access to the birthing areas.

Lyme Disease

Lyme disease is a zoonosis transmitted by ticks in the genus *Ixodes* (e.g., *Ixodes ricinus* in Europe or *Ixodes scapularis* and *Ixodes pacificus* in North America), which belong to the family of hard ticks, Ixodidae (see Figure 7). According to 2019 data from France's Sentinel Network, there are an estimated 50,000 new cases of Lyme disease in the French population each year. Figures for Europe remain poorly characterised — there may be thousands to hundreds of thousands of infections each year. This number is over 400,000 in the United States. The disease is caused by spirochetes and, more precisely, by pathogenic bacteria in the *Borrelia burgdorferi sensu lato* species group. This group currently has 20 described members, of which five are known to be pathogenic in humans. These pathogens appear to have evolved from commensal ancestors found in the ancestors of present-day ticks, before the hard tick lineage split off from soft tick lineage. Soft ticks (Argasidae family) also transmit *Borrelia* species that cause recurrent fever.

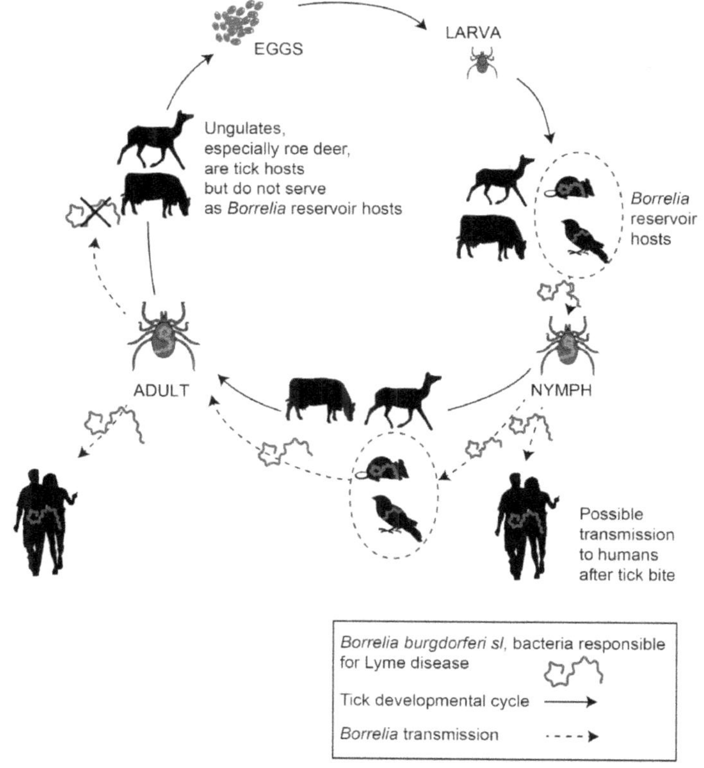

Figure 7. Lyme disease transmission cycle.

Lyme disease was first described in 1975 in the town of Lyme, Connecticut, USA. It had caused an arthritis epidemic in children. In 1982, a US researcher of Swiss origin, Willy Burgdorfer (1925–2014), isolated the bacterium. It was named after him and microbiologist Amédée Borrel (1867–1936), a student of Louis Pasteur. While the disease was described in 1975, it had certainly been in existence long before then. The disease's characteristic sign (erythema migrans) was first mentioned in 1905 by a Swedish physician (Afzelius). Then, in 1922, two French people (Garin and Bujadoux) provided the first description of neurological Lyme disease. Dating even further back, *B. burgdorferi* DNA sequences were found on a naturally mummified human (Ötzi), who was discovered in Italy and who lived 3,300 years ago.

Ixodes ticks have four developmental stages: eggs, larvae, nymphs, and adults. Larval and nymphal ticks feed on many types of vertebrates: mammals, birds, and scaled reptiles (i.e., squamates). Adults tend to be found on large mammals, especially ungulates, such as roe deer (*Capreolus capreolus*) in Europe or white-tailed deer (*Odocoileus virginianus*) in North America. Ticks become infected after feeding on infected animal reservoirs, mainly rodents and birds. Adult ticks seldom transmit the infection to their offspring, so larvae are rarely infected. Depending on the region, about 5–20% of nymphs and 15–25% of adults are infected. Different *Borrelia* species have different reservoirs. For example, in Europe, *B. afzelii* and *B. garinii* preferentially infect rodents and birds, respectively. In North America, *B. burgdorferi sensu stricto* infects both rodents and birds. Although deer serve as major hosts for adult ticks, they are not *Borrelia* reservoirs. Because ticks spend most of their lives in the outdoors rather than on hosts, they are very sensitive to changes in environmental conditions.

The bacteria are passed along to humans by infected adult or nymphal ticks. Generally, it takes the bacteria at least 24 hours to move from the tick's digestive tract (where the bacteria occur when ticks are not feeding) to the salivary glands (whence they can be injected). If a person's immune system does not destroy the pathogen, disease development occurs over three stages. The disease's exact clinical manifestations differ depending on the specific *Borrelia* species involved. The early localised phase is cutaneous: an erythema migrans appears. It is an expanding circular rash (a "bull's eye") that develops around the location of the tick bite in the subsequent month. The rash appears in 50–80% of cases (figures are highly approximate). It eventually disappears on its own. The disease symptoms may clear up by themselves, or the disease may move into its second stage, known as early disseminated Lyme disease. It brings with it a variety of symptoms that can be neurological, dermatological, rheumato-logical, cardiac, and ophthalmic. If no action is taken, the disease can enter the late disseminated stage, during which there are major effects on the neurological system, the skin, and the joints.

Diagnosis is based on a set of indicators that include physical symptoms, the results of biological tests, especially serological tests, and an evaluation of risk factors. Antibiotics are effective in treating the disease, especially early on. However, as the disease progresses, its symptoms may persist even after the infection has been brought under control with antibiotics. Ticks can also transmit other pathogens, such as tick-borne encephalitis virus or bacteria in the genera *Anaplasma* or *Rickettsia*. It remains unknown how coinfections might affect the disease's symptoms.

Since no vaccine currently exists, prevention largely rests on avoiding contact with ticks, removing any ticks from your body as soon as possible, and watching for symptoms afterwards so that swift action can be taken.

Plague

The plague is one of the few diseases to have left a clear mark on human history. It took several centuries for populations in the Old World to rebound from the mortality caused by the medieval plague epidemic, which was later referred to as the "Black Death". It is important to note that the term "plague" is frequently used in a variety of contexts in which it always denotes an event of major significance.

Plague is a disease caused by *Y. pestis*, a bacterium found in various rodent species that live in semi-arid areas. It is primarily transmitted via the rodents' fleas. The disease appears to have historically originated in China and parts of Central Asia.

Different rodents, including ground squirrels (family Sciuridae) and desert murids (genus *Meriones*, which includes gerbils), may have been the original hosts. These animals live in burrows, a habitat in which fleas and bacteria can survive for several months even after their hosts depart. Fleas lay eggs in the litter that serves as bedding, which is where the larvae develop. It is only after the fleas go through their last moult, reaching maturity, that they become haematophagous and colonise the fur of a host. It is hypothesised that black rats (*Rattus rattus*) passed along the bacteria to humans, after becoming commensal and being infected by a wild rodent species.

Historically, there have been three major plague pandemics. The earliest was the Plague of Justinian, named for the Byzantine Emperor during whose reign it occurred. The pandemic lasted from the mid-6th century to the mid-8th century and mainly spread around the Mediterranean Basin. It may have contributed to the fall of the Roman Empire. The Black Death began in the middle of the 14th century and lasted until the 18th century, ravaging Europe as well as parts of Asia. Europe lost at least 25% of its population. The third pandemic can confidently be dated back to 1894. It began in China, moved through India and Hong Kong, and eventually spread across most of the world. It established permanent footholds in Madagascar and the Americas, to cite two examples. At present, only Europe and perhaps a few islands and archipelagos appear to be plague free.

The dynamics of each pandemic are linked to the rodents involved in the pathogen's spread, namely the commensal behaviours of the black rat early on and then those of the Norway rat (*R. norvegicus*) in the 18th century. An inhabitant of Asia's warmer regions, the black rat likely arrived in Europe during the 1st century, as trade flourished, and the Silk Road came into being. However, the rat seems to have become particularly well established during the Middle Ages, which could explain why the Black Death had a greater geographical spread than the Justinian Plague. The Norway rat came from Asia's colder regions (e.g., China or Mongolia) and settled in Europe around 1700. The Norway rat is larger and more aggressive than the black rat and seems to have progressively supplanted it. The black rat disappeared entirely from some areas; in others, it took up residence in people's attics. In contrast, the Norway rat began occupying basements, cellars, and sewers. The Norway rat also appears to be more resistant to *Y. pestis* and hosts a different flea species, *Nosopsyllus fasciatus*. The flea found on the black rat is *Xenopsylla cheopis*. This difference could also partially explain the course taken by the most recent plague pandemic, recognising that the world changed dramatically between 18th century and the early 20th century.

When a black rat infected with the plague begins to die, its internal temperature drops, a signal picked up on by the resident fleas. They eat one last blood meal before heading off in search of a new host. If their host's blood is rich in plague bacteria, the latter form a plug in the anterior part of the flea's digestive tract. Only flea species in which solid plugs are formed are effective disease vectors. Such is not the case for *N. fasciatus*. However, for *X. cheopis*, the first step after moving to a new host (rat or human) is to attempt to expel the plug, which interferes with the flea's ability to eat and leads to starvation. The flea thus vomits up infectious bacteria, which pass directly into the new host's bloodstream, transmitting the disease. In humans, infection causes swelling in the lymph node closest to the bacteria's point of entry. Such enlargements are called buboes, resulting in the disease form known as the bubonic plague. In contrast, in septicaemic plague, the bacteria are found in the lungs, and the disease can have dramatic respiratory effects. Indeed, it can be transmitted directly among humans via airborne droplets. Flea bites are not required for its spread, such as in the case of bubonic plague. If the illness goes untreated, it swiftly leads to death, within just a day or two.

In North America, a specific type of plague has been observed in the domestic cats of people who live at interfaces with populations of potential rodent hosts. Cats appear to develop the disease after capturing infected rodents. Then, more importantly, they pass the illness on to their owners. These epidemiological dynamics regularly result in small plague outbreaks in the western United States.

Fortunately, we have access to antibiotics, which are highly effective against *Y. pestis* when administered in a timely manner.

Tuberculosis

Tuberculosis, alongside other bacterial diseases caused by *Mycobacterium* species, brings us closer to the full sense of a zoonosis: bidirectional pathogen swapping between humans and other animals. Indeed, the bacterium responsible for human tuberculosis, *M. tuberculosis*, appears to have evolved from a soilborne ancestor. It secondarily adapted to humans, becoming a strict parasite.

The shared history of *H. sapiens* and *M. tuberculosis* stretches back to ancient times, to at least 70,000 years ago and thus well before the Neolithic. It is even possible that their encounter occurred prior to the appearance of *H. sapiens*. It has been hypothesised that bone lesions observed on the remains of *H. erectus* (from 500,000 years ago) could have resulted from a tuberculosis infection, which would support the above idea. This long period of coevolution has led the bacterium to become dependent on humans. In contrast, other mycobacteria are primarily soilborne, sometimes secondarily acting as pathogens, sometimes not.

At present, tuberculosis is one of the most significant diseases afflicting humans. All causes combined, 10 million cases and 1.4 million deaths were reported worldwide in 2017. If the bacterium's impact is clearcut, its evolutionary history is quite convoluted. The genus is generally broken up into two groups: bacteria in the *Mycobacterium tuberculosis* complex (MTBC), a group that includes *M. tuberculosis* and *M. bovis* (agent of bovine tuberculosis), and the non-tuberculous mycobacteria, a group that includes *M. avium* (see p. 72) and *M. leprae*, the human-specific pathogen responsible for leprosy.

MTBC diversity is greatest in Africa, which fits well with our current understanding of human evolution. It would seem that *H. sapiens* and *M. tuberculosis* both arose in Africa and then left, likely together. There are also mycobacteria that are adapted to other mammal species, such as *M. orygis* in African antelopes, *M. mungi* and *M. suricattae* in African mongooses, as well as

M. bovis and *M. caprae* in domestic ruminants. Delving into the history of *M. bovis*, the infectious agent responsible for bovine tuberculosis, it appears that the pathogen moved from humans to cattle during the domestication process, perhaps 8,000 years ago. Other mycobacteria moved in the opposite direction. In pre-Columbian America, humans living along the Pacific coast likely experienced *M. pinnipedii* infections transmitted by seals and sea lions. European colonisation then brought *M. tuberculosis* to the New World, where it became established.

In brief, here are the histories of these three bacterial species:

- The human pathogen *M. tuberculosis* is not zoonotic. Rather, its ancestor was a soil bacterium that adapted to humans. It is sporadically passed along to bovines, dogs, and cats.
- The cattle pathogen *M. bovis* originally came from humans; its lineage diverged from that of *M. tuberculosis* around the time of domestication. *Mycobacterium bovis* is a zoonotic pathogen that can be transmitted to humans. In countries where milk is not routinely pasteurised, approximately 10% of human tuberculosis cases result from *M. bovis*. In some Western countries, bovine tuberculosis outbreaks are more of an economic concern than a public health concern. There are likely several factors at play: the growing number of larger livestock farms, increased animal trade during which health risks are poorly monitored, and an apparent decline in screening efforts. An unfortunate and unexpected consequence is that bovines have then infected several wildlife species, leading to the emergence of new epidemiological patterns. In Western Europe, three wild species then seem to boomerang the bacterium back to bovines, namely the wild boar (*Sus scrofa*), the red deer (*Cervus elaphus*), and the European badger (*Meles meles*).
- *M. pinnipedii* is specific to its pinniped hosts. However, since it can be transmitted to humans, it is a potential zoonotic agent. Hansen's bacillus (*M. leprae*) is the agent responsible for leprosy. It seems likely that its ecology better fit humans' ancient, pre-agricultural, and pre-industrial lifestyles. However, new questions have arisen over recent decades. In 2008, the bacterium was discovered to have another relative (*M. lepromatosis*), and, in 2016, both *M. leprae* and *M. lepromatosis* were found to co-occur

in populations of Eurasian red squirrels (*Sciurus vulgaris*). It appears that *M. lepromatosis* strains isolated from red squirrels in the UK diverged from *M. lepromatosis* strains in humans around 27,000 years ago. Conversely, the *M. leprae* strains found in UK red squirrels are most similar to *M. leprae* strains that were circulating in humans in medieval England. We are still waiting for a more in-depth examination of this discovery.

Antimicrobial Resistance

Antimicrobial resistance is not a zoonosis in the strict sense of the word because it is not a disease transmitted between humans and other animals. However, the fact is that microbial genes permitting resistance are swapped between animals and humans. For example, research has shown that the gut microbiota of pig farmers is influenced by the gut microbiota of their pigs. Consequently, antimicrobial resistance is increasingly being addressed as a zoonosis-related issue, as seen in European Directive 2003/99/EC on the monitoring of zoonoses and zoonotic agents. Resistance is a trait that all pathogens can display (e.g., bacteria, viruses, fungi, and parasites). Here, we will focus on antibiotic resistance.

The discovery of antibiotics was a revolution in the history of medicine. Used to treat bacterial infections, antibiotics transformed incurable diseases into easily cured illnesses. Deployed preventively, they allowed surgical practices to greatly advance by reducing the risks of infection during operations, and they helped protect vulnerable or immunocompromised people from opportunistic infections. Antibiotics are a cornerstone of modern medicine, and it is medically inconceivable to imagine life without them.

However, we are currently edging towards a future in which doctors can no longer properly treat infections caused by common bacterial pathogens. Indeed, more and more pathogenic or opportunistic bacteria are acquiring resistance to available antibiotics. These multidrug-resistant bacteria are particularly common in hospitals (i.e., regular and veterinary). In such settings, the selection pressure exerted by antibiotics is high, and numerous microorganisms coexist in proximity to a susceptible human population.

However, it is also worrying to witness the increasing presence of multidrug-resistant bacteria outside of hospitals.

That said, the problem of antibiotic resistance has existed for as long as antibiotics have been used for clinical purposes. By 1945, over 20% of the *Staphylococcus aureus* bacteria isolated in hospitals were resistant to penicillin, whose use in medical treatments had barely begun in 1942. Since then, the introduction of any new antibiotic has been quickly followed by the emergence of resistant bacterial pathogens. Variable mechanisms are involved, but they often rely on pump proteins that expel antibiotics from bacterial cells or enzymes that either inactivate the antibiotic or modify its target molecule. By 2050, antibiotic resistance is expected to become a leading cause of death in humans, resulting in an estimated 10 million or more deaths per year.

Antibiotics are used to prevent and treat bacterial infections in humans, other animals (e.g., ruminants, pigs, poultry, fish, or companion animals), and, more rarely, plants (e.g., horticultural species or fruit trees). Worldwide, 73% of the antimicrobials sold are used in animal-rearing operations. It should be noted that, aside from serving a veterinary purpose, antibiotics have been employed as feed additives. At low doses, they improve yields on pig, poultry, and cattle farms. The EU formally prohibited this use in 2006, the first of several proactive measures. In 2008, the EU implemented strong political incentives to reduce veterinary applications, which prompted declines in antibiotic use in animals, and, in 2018, usage levels in food-producing animals finally dipped below those in humans. However, such efforts have rarely been made in other parts of the world. In the US, for example, it was only in 2017 that the government banned using medically important antibiotics (i.e., those used in human medicine) to promote animal production and began requiring veterinary oversight for antibiotics administered via water or feed. Furthermore, these measures have yet to demonstrate their efficacity.

The massive deployment of antibiotics pollutes the natural environment. Indeed, humans and animals excrete active forms of antibiotic compounds, which then end up in wastewater even after treatment in sewage plants. In addition, a large portion of the antibiotics orally administered to livestock occur at high levels in manure; they contaminate the soil and groundwater when the manure is spread on fields. Finally, the pharmaceutical industry eliminates waste (e.g., active forms of antibiotic compounds) by directly discharging it into the environment, polluting aquatic habitats. Pollution can sometimes occur on a catastrophic scale, particularly in India and China, which produce most of the world's generic antibiotics.

Thus, resistant bacteria and their resistance genes are flowing among humans, other animals, and the environment. Outside hospitals, humans may be exposed to resistant pathogenic bacteria via contact with animals or the consumption of food contaminated with zoonotic bacteria, such as *Salmonella* or *Campylobacter*. The latter can become resistant following prolonged exposure to antibiotics in intensive livestock operations. In addition, animal-based foods can pass along resistant non-pathogenic bacteria capable of transmitting their resistance genes to members of the human digestive flora. This dynamic became evident when, in EU countries using avoparcin as a livestock growth promoter (a use since banned), enterococci emerged that were resistant to vancomycin, a closely related antibiotic. These bacteria were seen not only in production animals, but also in the intestinal flora of healthy humans and pets.

Resistant members of commensal flora occur in manure and sewage plant sludge, resulting in soil contamination upon spreading. Similarly, liquid waste from hospitals, livestock farms, and municipal sources contains bacteria carrying resistance genes, which can end up in aquatic habitats. In Paris, for example, non-enteric bacteria found in the Seine are resistant to quinolones. As a general rule, wastewater treatment plants and their effluent contain massive quantities of antibiotic-resistant bacteria in the family Enterobacteriaceae. It is difficult to assess the environmental spread of resistant bacteria because so many factors are at play. However, analyses have detected the presence of resistant bacteria in the commensal digestive flora of wild mammals and birds, revealing that the contamination is broad reaching. Wild animals may be more than just reservoirs. They may also disseminate resistant bacteria as they move around.

A 2022 report released by the UN Environment Programme underscored that the environmental dimensions of antimicrobial resistance are characterised by cyclic interrelationships and their complexities as well as by multiple causalities and dynamics. Pollutant releases, effluent, and waste arising from animal production play a major role in the above. Thus, it is important to adopt a systems approach, such as "One Health" (see sidebar p. 122), if we wish to better understand the environmental dimensions of microbial resistance and scientifically inform policy.

ZOONOTIC PROTOZOA AND WORMS

Echinococcosis

The genus *Echinococcus* contains very small tapeworm species. They measure just a few millimetres long and live in the intestines of their hosts — a variety of meat-eating mammals, notably domestic dogs. These cestodes (i.e., parasitic flatworms) have a two-host life cycle. Their eggs hatch into infectious larvae in one host and are then ingested by another host, in which they reach maturity and reproduce. The former species is called the intermediate host, and the latter species is called the definitive host. Western Europe is home to two *Echinococcus* species of public health concern: *Echinococcus granulosus*, which causes cystic echinococcosis, and *E. multilocularis*, responsible for alveolar echinococcosis. *Echinococcus granulosus* generally occurs in the Mediterranean part of Europe and lives mainly in dogs (its definitive hosts) and domestic ruminants (its intermediate hosts). *Echinococcus multilocularis* is found in continental Europe and lives in dogs and foxes (its definitive hosts) and various species of field voles (its intermediate hosts; certain species in the genera *Microtus* and *Arvicola*). In *E. granulosus*, humans replace ruminants as the intermediate hosts. In *E. multilocularis*, humans replace the voles.

For humans to be parasitised, they must ingest a sufficient quantity of tapeworm eggs from dogs or foxes. Such can potentially result from poor household hygiene; allowing dogs to have access to dishes; handling live or freshly killed foxes with one's bare hands; or eating contaminated fruits and vegetables. The larvae of *E. multilocularis* can cause debilitating lesions as they spread throughout various tissues (i.e., mainly the liver and sometimes the lungs). The infestation process is very slow, and the worm's presence can go unnoticed for several months or even years. In the absence of treatment, death can result. Fortunately, safer chemical treatments have replaced riskier surgical treatments.

Dogs become infected with *E. granulosus* after consuming goat or sheep viscera. The parasite forms what may be referred to as "water balls", hydatid cysts that are attached to various organs. The tapeworm is transmitted from its intermediate to definitive host when hygiene during slaughtering is poor or because of contamination outside of slaughterhouses. For *E. multilocularis*, dogs or foxes only become infected after eating voles. Even though tapeworm infestations are common in foxes in some regions of Europe, encounters with infested voles are far rarer. More than anything, this result means that foxes are much more skilled at catching voles than are parasitologists, although the latter may be less motivated hunters.

Historically, *E. multilocularis* has tended to occur in Central Europe. Since the early 21st century, its distribution seems either to be expanding or, more likely, is better characterised. A few foxes infested with tapeworms have been found far to the west of this traditional range. While it is possible that the parasite has been heading westward, it is also possible that the parasite gained in visibility following the strong recovery of fox populations after the decades-long rabies outbreak (1968–1998). Infestations may have been less perceptible when fox densities were lower. That being said, since infested foxes have been found in areas that were unaffected by rabies, both explanations, and as well as others, are certainly conceivable.

In areas where the tapeworm is endemic, it is recommended that dog owners prevent their animals from hunting voles. Furthermore, dogs should be regularly dewormed. Ultimately, the safest policy is to stay away from foxes. Indeed, the best management approach appears to be to "do nothing" in relation to foxes. This conclusion has been reached based on research looking at the impact of controlling fox populations or baiting foxes with foods containing anthelmintics.

Toxoplasmosis

Toxoplasmosis is a disease caused by *Toxoplasma gondii*, a single-celled parasite (i.e., protozoan parasite) that is found in a wide variety of mammals, including humans. It was described in the very early 20th century by Charles Nicolle, winner of the 1928 Nobel Prize for Medicine, who was director of the Tunis Pasteur Institute at the time. The parasite was discovered in a North African rodent species (family Ctenodactylidae) locally referred to as a "gundi", hence the protozoan's name.

Fortunately, clinical disease remains rare. At greatest risk are those who are pregnant or immunocompromised. During pregnancy, there is a risk of developing congenital toxoplasmosis — the parasite infects the foetus, potentially causing severe problems. In Europe, 208 cases of congenital toxoplasmosis were reported in 2018, 73% of them in France, which has an active prenatal screening programme. Starting with the AIDS epidemic, severe forms of the disease began to appear in adults. Notably, people would develop cerebral toxoplasmosis, where the parasite would localise itself in the brain. Such dynamics had been rare prior to that point.

Although many mammals and birds host the parasite, it would appear that sexual reproduction is only possible in felids. Humans mainly become infected after consuming raw or undercooked ruminant meat. More rarely, people eat fruits or vegetables contaminated by felid faeces, which contains oocysts that can remain viable even after more than a year outdoors. If a household contains someone at risk of infection and a cat that goes outdoors, precautions should be taken when cleaning the litter box. Ideally, the faeces should be disposed of immediately, as the oocytes sporulate and become infectious once outside the cat's body. Strictly indoor cats and cats fed commercially produced foods (e.g., canned wet food or kibble) do not present a disease risk.

The pathogen displays unique epidemiological dynamics in the Neotropics, such as in French Guiana, where no less than six wild felid species coexist (the jaguar, *Panthera onca*; the puma, *Puma concolor*; the ocelot, *Leopardus pardalis*; the margay, *Leopardus wiedii*; the oncilla, *Leopardus tigrinus*, and the jaguarondi, *Herpailurus yagouaroundi*). In contrast, only three felid species are found in Europe (the Eurasian lynx, *Lynx lynx*; the Iberian lynx,

L. pardinus, and the wildcat, *Felis silvestris*), although the EU is home to more than 75 million domestic cats. In the Neotropics, toxoplasmosis is transmitted via oocysts, which appear to be more common in this region than in other parts of the world.

Trichinellosis

Trichinella spiralis is a nematode with a particular lifestyle: it has no free-living stage and must always remain inside a host. Its larvae develop in the muscle tissue of carnivorous or omnivorous mammals. They transition out of latency when this muscle is consumed by another carnivorous or omnivorous mammal that is either a predator or a scavenger. In the new host's digestive tract, the larvae moult, become adults, and reproduce. The next generation of larvae traverse the intestinal barrier, spread through the body, and eventually settle down in the muscle mass that will house them until the cycle repeats itself. The parasite was far more common in the past, when pigs had access to the outdoors or when rats, which host the parasite, were numerous in pig-rearing facilities. Industrial farming has resulted in different conditions.

European Union regulations require that tests be conducted to determine whether the parasite is present before the sale of any "at-risk" meats (i.e., pork, wild boar, and horse). To this end, muscle samples are collected at slaughterhouses or game-handling establishments. For hunters who consume their own game, freezing in household appliances is not enough to properly treat the meat, since some trichinae are cold resistant. However, heat works quite well: larvae can be killed in around 3 minutes at 58°C and almost instantly at 63°C. These temperatures are easily reached when meat is thoroughly cooked. In humans, the disease can be quite painful when the larvae are migrating through the body. However, it is easy to treat once diagnosed.

A surprising case occurred in early 2010 in Corsica. A health inspection of pig carcasses was being conducted at a slaughterhouse. The animals had come from a modern pig-rearing operation. During the investigation, it was discovered that the farmer had fed his pigs the corpse of one of his dogs!

In the 1980s and 1990s, horse meat caused several major instances of trichinellosis in Italy and France. These events were also surprising because the parasite is only transmitted through the consumption of infected meat. It has no free-living stage. Given that horses are strictly herbivorous, the infection must have somehow occurred by accident. These cases remain a mystery.

VIRAL ZOONOSES

COVID-19 and Other Zoonotic Coronaviruses

At the beginning of the 21st century, veterinarians were perhaps the people most familiar with coronaviruses because they cause major diseases in livestock and affect companion animals. However, they were not as familiar to medical doctors. For example, the coronavirus family is responsible for transmissible gastroenteritis in pigs, infectious bronchitis in turkeys, infectious peritonitis in cats, and winter colds in humans (e.g., HCoV-229E or HCoV-OC43). Given the lack of common names for the two human cold-causing coronaviruses, it is clear that they are only of modest clinical and epidemiological importance. In 2002, the world's perspective on coronaviruses shifted with the emergence of severe acute respiratory syndrome (SARS) in the southern Chinese province of Guangdong. The epidemic of SARS-Cov-1, so renamed after the emergence of SARS-CoV-2, officially lasted from November 2002 to July 2003. Around 8,400 cases occurred in Asia, Europe, and America, and the disease resulted in nearly 900 deaths. The case fatality rate was therefore nearly 10%. SARS-Cov-1 had never been seen before and has not reappeared since.

Over the past two decades, researchers have considerably expanded understanding of the coronavirus family, which contains four genera: *Alpha-*, *Beta-*, *Gamma-*, and *Deltacoronavirus*. The first two comprise mammalian viruses, and the latter two comprise bird viruses. SARS-CoV-1 is in the genus *Betacoronavirus*, like MERS-CoV and SARS-CoV-2.

It appears that the emergence of SARS-CoV-1 was linked to human consumption of a small tree-dwelling primarily frugivorous omnivore, the masked palm civet (*Paguma larvata*; family Viverridae). This species is commonly eaten in southern China in particular. Indeed, the epidemic had its roots in restaurants where this animal was on the menu. In China, small animals are sold live, in markets and restaurants, to guarantee their freshness to consumers. The butchers working at these restaurants were the first to be affected. The virus' source was not traced back to hunters, breeders, vendors, or consumers of masked palm civets. The epidemic began spreading from person to person, and many of those afflicted infected their caregivers and/or other people with whom they came in contact. For example, a Chinese doctor travelled to Hong Kong on vacation after a long stint caring for patients in Guangdong, not knowing that he had also become infected. He stayed at an international hotel, where he then infected several businesspeople from various continents. They returned home with the virus. The epidemic's global spread is largely due to this doctor, who acted as an epidemiological superspreader. This man ended up hospitalised himself before ultimately passing away.

On the animal end of the equation, it is not certain that the masked palm civet is the virus' reservoir, even if it served to link wildlife and humans. Indeed, the strains found in humans have all differed from the strains found in civets. Subsequent research explored the viruses found in the region's markets, farms, and free-ranging wildlife and identified a group of coronaviruses similar to the strains responsible for SARS-CoV-1. They occurred in horseshoe bats, small bat species in the genus *Rhinolophus*, and shared a common ancestor with the viruses isolated from civets and humans. SARS-CoV-1 thus seems to have originated in bats. Two questions subsequently arise: First, how did a bat virus get passed to civets and then humans? Second, how did it evolve to become pathogenic in humans given that the related viruses do not appear to be pathogenic in either horseshoe bats or civets? These questions remain unanswered.

Another surprise was in store: the 2012 discovery of a *Betacoronavirus*, MERS-CoV, on the Arabian Peninsula. Transmission is ongoing. As of late 2021, a total of 2,578 cases and 888 deaths had been reported. The epidemiological dynamics seem comparable: a viral ancestor in Asian bats (genus *Taphozous*) and a terrestrial mammal, the dromedary *(Camelus dromedarius)*, acting as the source of human infections. However, transmission among humans is much less efficient than for SARS-CoV-1. New human cases have always been linked to contact with infectious dromedaries. Serological surveys carried out in all the world regions in which dromedaries are kept, from the Canary Islands to Central Asia, have revealed the omnipresence of antibodies in dromedaries. So why did the virus only emerge in humans in 2012 in the countries found on the Arabian Peninsula? There are no answers to these questions yet either.

Finally, in late 2019, news out of central China revealed the presence of an apparently contagious and transmissible respiratory illness of as yet unknown aetiology. It immediately evoked memories of the SARS-CoV-1 epidemic for those who had been involved. Unfortunately, it indeed turned out to be another member of *Betacoronavirus* — SARS-CoV-2 — that was also a close yet distinct relative of the viruses found in Asian horseshoe bats. Seventeen years after the SARS-CoV-1 epidemic, this new disease, named COVID-19 (for *COronaVIrus Disease 2019*) caused a new epidemic that rapidly turned into a pandemic. While SARS-CoV-2 is far more transmissible than SARS-CoV-1, it fortunately has a much lower case fatality rate ($< 1\%$ and probably closer to 0.5%). Even now, in late 2021, it remains far too early to answer all the outstanding questions regarding the emergence of COVID-19. Initial comparisons of strains from bats and humans suggest that the "humanised" virus had been circulating for several years before it emerged in a medically identifiable way. Did a terrestrial mammal play an epidemiological role in this situation? Initially, researchers stumbled upon a coronavirus in some smuggled Malayan pangolins (*Manis javanica*). However, this discovery remains tricky to interpret and predates the beginning of the COVID-19 pandemic. Pangolins are sold in markets where they are crowded together with dozens of other species under very unsanitary conditions. These circumstances are conducive to the interspecific transmission of microorganisms, including to humans. Some people have also hypothesised that the virus accidentally escaped from a high-level biosafety laboratory. Such has occurred in the past. For instance, the virus responsible for foot-and-mouth disease leaked out of a UK laboratory in 2007, and SARS-CoV-1 escaped from Taiwanese and Chinese laboratories between 2003 and 2004. However, there was no evidence to support this hypothesis for SARS-CoV-2 at the time this book was written. From a broader perspective, we lack virus sequences that are sufficiently related to SARS-CoV-2 across both time (i.e., for a retrospective analysis) and space (i.e., for a geographical analysis); such would be needed to truly identify the virus' origin and emergence mechanisms.

Based on data from Europe and the rest of the world, it is known that SARS-CoV-2 can be transmitted to domestic animals or wild animals. Starting in 2020, small numbers of infections were observed in pets (e.g., dogs, cats, golden hamsters); farm animals; zoo animals; and wild animals. Felids and mustelids have been particularly affected. Some farmed American minks (*Mustela vison*) have been contaminated by facility staff. Furthermore, it has been reported that the virus can move back from minks to humans. In Europe, these infections resulted in the slaughter of all the minks on farms where the virus was detected. Those on nearby farms were also put down as a preventive measure. In North America, people transmitted the virus to white-tailed deer (*Odocoileus virginianus*), which may have reinfected humans in turn.

Other, highly diverse members of *Betacoronavirus* continue to circulate in wildlife. Given that we have witnessed the emergence of three coronaviruses to date, we should not underestimate the possibility that it will happen again.

Crimean-Congo Haemorrhagic Fever

Crimean-Congo haemorrhagic fever was first described in 1944 among Soviet soldiers stationed in Crimea. Its viral agent was isolated in 1956 in the Congo. One of 25 viruses that can cause haemorrhagic fever, it is an RNA virus in the genus *Orthonairo-virus* (family *Nairoviridae*, order *Bunyavirales*). Like the virus, the virus' genus was named for a geographical location: Nairobi.

The virus circulates via an enzootic cycle, from ticks to vertebrates back to ticks. Tick bites are the cause of most transmission events, but humans can also become infected through contact with bodily fluids. Crimean-Congo haemorrhagic fever has a distribution that broadly matches that of its main vectors, namely ticks in the genus *Hyalomma*. These ticks are easily identifiable due to their white- and yellow-striped legs. The disease may expand its range even further as a result of global warming, the introduction of *Hyalomma* into new areas by migratory birds, and the international livestock trade. In Europe, *H. marginatum* is spreading throughout the Mediterranean. Unlike its relative *Ixodes ricinus*, this species likes dry climates, such as those found in scrublands and dry hills. *Hyalomma marginatum* also differs from *I. ricinus* in that it actively seeks out hosts. It does not employ a sit-and-wait strategy. Instead, it detects potential hosts via olfactory or vibratory cues and can quickly travel several meters to intercept them.

The tick's larvae and nymphs mainly occur on small mammals (especially lagomorphs) and birds (mainly songbirds). The adults are generally found on large mammals, including horses, cows, wild boars, and deer. While the virus circulates in the blood of its vertebrate reservoirs for only short periods, it remains in its tick vectors over their entire lifetimes and is even passed along to their offspring.

To date, Crimean-Congo haemorrhagic fever has been observed in eastern and southern Europe, the eastern Mediterranean Basin, northwestern China, Central Asia, the Middle East, and several African countries. It is also thought to be circulating in North Africa, but concrete evidence for this hypothesis is lacking. The various cases reported in Europe have mostly occurred in Spain and Bulgaria.

In humans, some cases likely go undetected, but the case fatality rate is high (10–40%). Treatments mainly aim to relieve symptoms, but an antiviral drug can be administered as well if needed. A vaccine exists and was first used in the 1970s in the former USSR. However, it elicits an imperfect immune response. It remains challenging to control animal and tick populations. Therefore, in areas where the virus is endemic, prevention is the main approach, namely avoiding tick bites and exposure to the blood and bodily fluids of infected animals and humans.

Yellow Fever

The history of yellow fever is intimately intertwined with that of humans in rather tragic ways. It is a disease caused by members of the genus *Flavivirus* (family Flaviviridae), which are transmitted by different mosquito species that only occur in the tropical regions of Africa and the Americas (see Figure 8). The first clinical descriptions of yellow fever come from the Americas, where the virus and the disease were introduced as early as the 16th century because of European chattel slavery, in which Africans were part of the sinister Triangular Trade system. For a long time, Europeans limited themselves to exploring the coastline of Africa, without heading towards the interior, which is where yellow fever occurs.

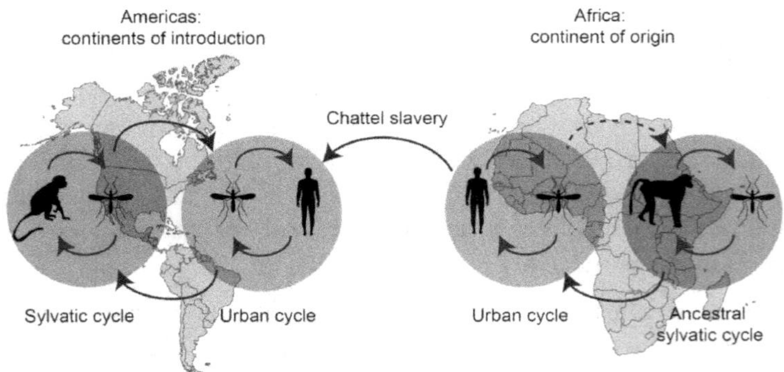

Figure 8. Transmission of yellow fever virus in Africa and the Americas.

Considering the conditions associated with those past transatlantic crossings, it seems likely that the virus was carried not by sick humans, but rather by mosquito eggs. Indeed, infected mosquitoes (genus *Aedes*) can transmit the virus to their eggs, which are highly resistant. In the 19th century, a crucial epidemiological discovery was made: the involvement of mosquitoes in the disease's transmission, a finding that was also an essential step in our current understanding of diseases, particularly those caused by arboviruses. Up until that point, it was thought that yellow fever spread via airborne transmission among people. Interestingly, local mosquitoes were immediately able to vector the virus upon its arrival in the Americas, even though they had never encountered it before. The *Aedes* species established themselves in the areas where humans settled. However, the mosquito species native to South American forests (genera *Haemagogus* and *Sabethes*) quickly took over vectoring the pathogen.

Seven yellow fever virus genotypes have been described to date. They have specific geographical associations — five occur in Africa and two in the Americas. The two genotypes in the Americas are clearly similar to the genotypes in West Africa, reflecting the disease's transatlantic history. Molecular comparisons have shown that the strains all originated in Africa and that the strains that ultimately became specific to South America diverged around the middle of the second millennium CE. As for the original yellow fever virus, it seems to have evolved 3,000 years ago somewhere in Africa from a *Flavivirus* ancestor.

While non-human primates in Africa can be infected by the virus, it does not appear to greatly affect them as they do not manifest any clinical signs of disease. The situation is different in the Americas, where human epidemics often follow the dramatic waves of mortality among native primates. Thus, residents and visitors alike are asked to avoid affected areas entirely or only travel there if properly vaccinated. Clearly, the epidemiological pattern in the Americas suggests that monkeys are not reservoirs. Instead, this role is played by mosquitoes. Similar questions have been raised regarding the virus' dynamics in Africa.

There is a sylvatic viral cycle that involves monkeys and forest mosquito species. These same insects can infect humans that enter the forest or that live in villages near the forest. There is also potentially an urban viral cycle. Other mosquito species (genus *Aedes*) are anthropophilic and live in cities. For instance, an urban transmission cycle can develop if someone who is viraemic (i.e., carrying a viral load in their blood) arrives in the city from the forest and is bitten by *A. aegypti*. Such dynamics have been observed in several large US cities, found well north of the tropics. Up until the end of the 19th century, epidemics took place in major port cities in the eastern US, causing great waves of illness and death. The same occurred in Barcelona, Spain (1821–1822). However, it seems that the temperate climates of both the Spanish and eastern North American coastlines prevented the virus from firmly establishing itself.

One issue that continues to puzzle epidemiologists is that yellow fever epidemics have never been observed in Asia, even though trade between Asia and Africa has multifarious and ancient roots. There may be several factors at play. First, Asian insects could be poor vectors for East African virus strains. Second, there might be competition between two *Aedes* species, namely *A. aegypti* and *A. albopictus*. *Aedes aegypti* arrived in Asia by crossing the Atlantic Ocean, the Americas, and then the Pacific Ocean. It did not come more directly from Africa. Third, humans living in Asia might have cross-protective immunity because of exposure to more local *Flavivirus* species. Notably, dengue virus has been in circulation for a long time in that part of the world.

One French overseas department, French Guiana, is located in an endemic zone, and vaccination against yellow fever has been required there since 1967. Even so, since 2017, three cases have occurred as a result of local transmission, including one with a fatal outcome in July 2020.

Influenza

In humans, influenza is caused by viruses in the family Orthomyxoviridae and the genus *Influenzavirus*. They commonly occur in humans and in various species of birds and mammals (see Figure 9). There are four antigenic types: A, B, C, and D. Types A and B cause seasonal influenza epidemics, and only type A viruses have led to pandemics to date. Type C viruses cause sporadic cases of influenza, while type D viruses, found in pigs and ruminants, are not considered to be pathogenic to humans.

Here, we will focus on influenza A viruses because they are commonly encountered in both humans and various animal species. These viruses can evolve quite rapidly by exchanging segments of genomic RNA. Each virus is composed of eight separate segments (see Figure 10). Influenza A viruses are divided into subtypes based on the combination of two proteins expressed on the surface of the viral envelope: haemagglutinin (H) and neuraminidase (N). Consequently, subtype names take the form

HxNy, where there are 18 haemagglutinin and 11 neuraminidase proteins described to date (i.e., H1 through H18 and N1 to N11, respectively). Almost all the subtypes occur in wild waterfowl, which are most likely the natural reservoir for the influenza A viruses found in other animal species, including humans. However, new influenza subtypes (H10N17 and H11N18) have been discovered in bats, raising questions about their possible role in the ecology of influenza A viruses. To date, only viruses carrying haemagglutinin types H1, H2, or H3 and neuraminidase types N1 or N2 have adapted to humans. They cause the "human flu", which is characterised by highly effective human-to-human transmission. Humans can also sporadically contract influenza viruses from birds or pigs, causing cases of "zoonotic influenza". Influenza viruses found in horses, dogs, and cats are generally not zoonotic, although incidents of transmission have been described, particularly from humans to dogs.

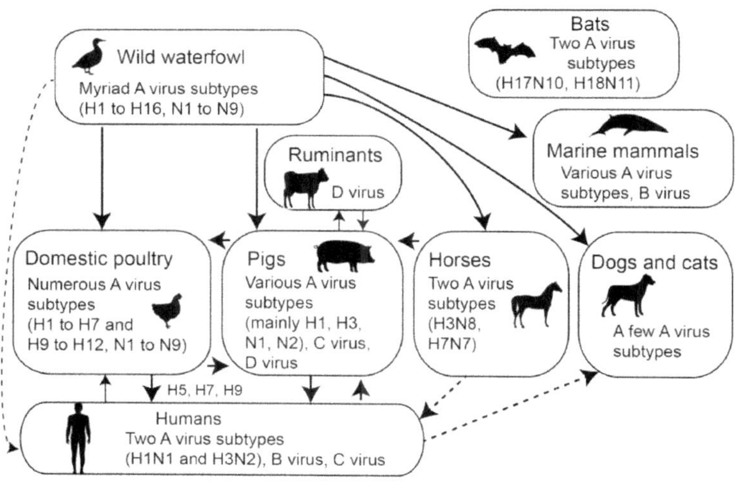

Figure 9. Influenza viruses in different host species.

The main types and subtypes of influenza viruses circulating in different taxonomic groups are mentioned in each box. The arrows show the interspecific exchanges described in the text (dotted line: rare exchanges). Information taken from Mourez T., Burrel S., Boutolleau D., Pillet S., 2019. *Traité de virologie médicale*, Edition Société française de microbiologie.

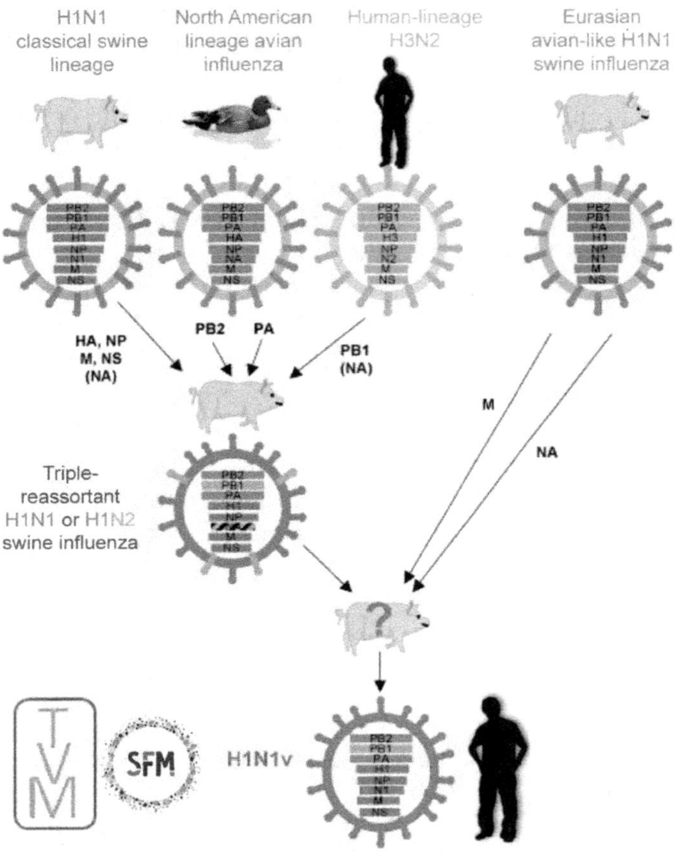

Figure 10. Origin of the influenza A(H1N1)pdm09 virus.

The NA (neuraminidase) and M segments came from a porcine virus with a Eurasian lineage. The other segments came from a triply assorted swine virus: A(H1N1) or A(H1N2). Information taken from Mourez T., Burrel S., Boutolleau D., Pillet S., 2019. *Traité de Virologie Médicale*, Edition Société Française de Microbiologie and based on Munier *et al.*, 2010.

Doomestic poultry can be infected with a wide variety of influenza A viruses, which they then shed in their droppings. With some viruses, the animals may not show clinical signs of infection (i.e., only detectable via laboratory testing). Alternatively, their symptoms may be relatively mild: decreased appetite, decreased egg laying, and/or light respiratory problems, such as spitting or coughing. In contrast, other viruses may be extremely pathogenic, sometimes causing what is referred to as "fowl plague". In such cases, the animals experience severe respiratory, digestive, or nervous system damage. There is also sudden and massive mortality. The emergence of these extremely transmissible "highly pathogenic" strains in poultry is the consequence of industrial farming conditions, in which young birds of the same genetic background are packed together at elevated densities. The transmission of highly pathogenic influenza viruses to wild birds can also have serious consequences for biodiversity and endangered species. Some subtypes (to date: those containing H5, H7, H9, or H10) can end up in humans via the inhalation or mucosal carriage (i.e., via the hands) of virus particles on bird plumage. Most of these infections simply cause mild conjunctivitis or transient respiratory problems, but others, especially those involving subtypes H5 and H7N9, can impact the lower respiratory tract and lead to death. The H5N1 virus is responsible for the "bird flu" that has been circulating in Asia since 2003. It was present in western Europe in the winter of 2005–2006. In spring 2020, an overall assessment of the virus' effects revealed its very high case fatality rate in humans (63%): out of the 861 cases identified since 2003, 455 had resulted in death. Fortunately, the rate of infection has remained low.

Pigs can be infected with various influenza A subtypes, predominantly H1N1, H1N2, and H3N2, and with human and avian influenza viruses. Swine viruses can acquire genes from avian and human viruses via the exchange of genomic segments, a process known as reassortment. They thus serve as a type of unique host, one that facilitates the emergence of new subtypes to which the human population has no immunity. These new viruses can then move freely across the globe, causing an influenza pandemic. This situation occurred in 2009 when the A(H1N1)pdm09 virus emerged and spread worldwide within a few weeks. It was unrelated to the H1N1 seasonal influenza viruses that had been circulating in human populations since 1977. European countries launched emergency vaccination campaigns, which were accepted to varying degrees by their populations. In the end, vaccination proved to be unnecessary because the virus caused less severe disease than feared. However, when faced with a highly virulent virus that spreads rapidly via a respiratory route, vaccination is the most effective means of protecting public safety, provided that this tool is available. It is therefore essential to monitor circulating swine influenza viruses so that researchers can prepare to develop any necessary vaccines ahead of time. For example, in 2020, an alert was issued regarding a new multiply reassorted virus in pigs, called Genotype 4 Reassortant Eurasian Avian-like H1N1. Two cases of transmission to humans were reported in China, in 2016 and 2018, but no documented human-to-human transmission occurred in either case. These sporadic outbreaks highlight the importance of implementing strict biosecurity measures on pig farms to limit the risk of influenza viruses passing between humans and pigs.

West Nile Fever

West Nile virus is named for the place it was first isolated in 1937: the West Nile subregion of Uganda. It is a member of the genus *Flavivirus*, which also includes the dengue and yellow fever viruses. Like the latter, West Nile virus is an arbovirus. Its natural transmission cycle involves wild birds, which serve as amplifier hosts, and ornithophilic mosquitoes, which serve as vectors. The virus can also be transmitted to mammals by mosquitoes that have previously fed on infected, viraemic birds (i.e., with large quantities of virus in the blood). Humans and horses are two of the susceptible mammal species that can develop symptoms ranging from simple fever to severe encephalitis. That said, most infections are asymptomatic.

Until the late 1990s, West Nile virus had only been detected in the Old World, mainly in Africa, the Middle East, and Europe. In 1999, the virus suddenly appeared around New York for a reason that remains unknown. It caused clinical disease in hundreds of people and equines. High mortality rates were seen in zoo birds and wild birds, particularly corvids. The virus then rapidly spread across North America. Birds served as sentinels, providing early warning of the disease's arrival. It appeared in 2001 in Canada; in 2002 in the western US, Mexico, and the

Caribbean; and in 2006 in South America. Wild birds probably played an important role in its spread.

West Nile virus' epidemiological dynamics in the Americas are probably related to two factors: first, the introduced strain seems to have been particularly virulent and, second, its spread occurred within fully susceptible bird populations (i.e., ones that did not coevolve with the virus). In Europe and the Mediterranean Basin, the epidemiological situation is different. While the virus was first identified in this geographical region in the 1950s and 1960s, it has probably been in circulation for far longer. Nonetheless, the disease is considered to be re-emerging even there because West Nile epidemics and epizootics have become increasingly frequent since 1994. Could it simply be that monitoring efforts have improved?

In the Camargue region of France, a West Nile epizootic occurred in equines in the late summer of 2000, after a seeming absence of more than 30 years. A total of 76 clinical cases were confirmed; one-third of the animals died because of the disease or were euthanised. Since then, West Nile virus circulation has been reported several times in French administrative departments around the Mediterranean. These events have affected horses and/or humans. West Nile virus was isolated from the brains of wild birds in 2004 and 2018. However, unlike in the U.S., no abnormal mortality has been seen in French bird populations. In southern France, West Nile monitoring efforts comprise four complementary facets that focus on vectors, birds, horses, and humans, respectively. As a result, virus circulation is detected early on, allowing the rapid implementation of preventive and protective measures, mainly to ensure the safety of blood donations and organ transplants.

Ebola

Ebola viruses are named after a river in the northern Democratic Republic of the Congo (DRC), the region where one of the first known human cases occurred in 1976. These viruses cause haemorrhagic fever, characterised by the sudden onset of fever, intense fatigue, headaches, and muscle pain. The latter symptoms are often followed by digestive disorders. There may also be signs of cutaneous and mucosal haemorrhaging. The case fatality rate is 25–90%, depending on the epidemic and the virus species. Between 1976 and 2014, about twenty epidemics occurred in isolated regions of Central Africa (i.e., in DRC, Sudan, Uganda, and Gabon). Then, between 2014 and 2016, an epidemic of unprecedented proportions occurred in West Africa, a region that had largely been spared, with the exception of an isolated case in 1994. More than 20,000 people died. The US and some European countries (Spain, Italy, and the United Kingdom) have been sporadically affected due to travellers from these affected regions developing disease symptoms upon arrival. Internationally coordinated control campaigns have managed to reduce transmission. In 2018, another outbreak occurred in the eastern DRC. In 2020, it was followed by another in the western DRC.

Outbreaks in human populations often occur subsequent to unusual mortality events in great ape populations. The apes show symptoms similar to those seen in humans and are thought to become infected via an animal reservoir. Once in humans, the virus is spread through direct contact with the blood, secretions, or other biological fluids (e.g., saliva, sweat, semen, vomit, and/or faeces) of those infected. Transmission largely takes place among the family and health care personnel who are looking after the sick person. Consequently, methods for preventing human-to-human transmission include single-use equipment, patient isolation, and zero contact with the infected, even after death. In 2015, a vaccine (VSV-ZEBOV) was developed that is administered during outbreaks.

Ebola viruses have been found to circulate in animal species other than primates, notably in the frugivorous flying foxes (family Pteropodidae). They are suspected to be the natural reservoir for Ebola viruses in Africa as well as on other continents. However, to date, we have no clear evidence of the virological role played by bats. More research is needed to better understand Ebola virus diversity in wild reservoirs and pathogenicity in humans. Indeed, some Ebola viruses can infect humans without causing disease, including the Ebola Reston virus, found in macaques and pigs in the Philippines.

Rabies

Humans seem to have had historical knowledge of rabies in dogs and wolves (the latter is a wild version of the former). However, the disease's causative agent, the rabies virus, was not identified until the early 20th century. The disease has been observed in different members of Carnivora found in multiple parts of the world at varying points in time.

From 1968 to 1998, rabies was rampant in red foxes (*V. vulpes*) in France. This epizootic has been well studied (see sidebar p. 134). Its origin appears to trace back to Central Europe, possibly Poland, and the 1930s or 1940s. A canine virus strain seems to have adapted to the red fox. Indeed, vulpine rabies had not been previously described, or only in anecdotal terms. In all species, the virus has an incubation period of several weeks to months. Then, the emergence of clinical disease is triggered. This phase lasts a few days and always ends in death. Consequently, for rabies, it is important to recognise that the reservoir is maintained at the population level rather than at the individual level. In their various mammal hosts, virus strains provoke an adaptive pattern of transmission. Infected individuals shed virus in their saliva during the clinical phase, or even a few days prior. They also display a behavioural shift that promotes the virus' spread before host death occurs. To take the example of the red fox, sick animals would come out in broad daylight and move around in seemingly random patterns. This behaviour would attract the attention of healthy foxes. Foxes or fox families establish fairly exclusive home ranges and delimit their boundaries with scent marks. If these markings are not regularly renewed and if a territory's residents are wandering around aimlessly, the neighbours will come out to see what is happening. They will thus encounter the rabid fox, get bitten, and become infected.

An epizootic wave can eliminate up to 90% of the local fox population, which will take an average of three to four years to recover. This time frame is relatively quick for a species with a single breeding season and a single annual litter. However, a large number of survivors reproduce, litter size is above average, juvenile survival is improved, and juveniles breed earlier. This pattern explains why culling-based strategies have failed to control fox rabies.

The rabies virus (genus *Lyssavirus*, family Rhabdoviridae) has long been considered a classical example of a monotypic virus. Historically, only one strain (now called RABV) had been described. Over the course of the 20th century, three slightly different virus species were discovered, all of African origin. They are Lagos bat virus (LBV), Mokola virus (MOKV), and Duvenhage virus (DUV). The reservoirs for LBV and DUV are bats. In contrast, that of MOKV is still unknown; it has been isolated from various terrestrial mammals. From the 1980s onwards, a series of new *Lyssavirus* species were discovered, all but one occurring in bats. Several occur in Europe: *European Bat Lyssavirus 1* and *2* (EBL1 and EBL2), *Bokeloh Bat Lyssavirus* (BBLV), and *Lleida Bat Lyssavirus* (LLEBV). Each species of virus appears to be associated with a particular species of bat. For instance, in France, EBL1 has been found in the common serotine bat (*Eptesicus serotinus*) in 1989; BBLV in Natterer's bat (*Myotis nattereri*) in 2012 and 2013; and LLEBV in Schreibers' bent-winged bat (*Miniopterus schreibersii*) in 2017. While EBL1 seems to be around every year, the three other species appear far more rarely. EBL2 has never been observed in France but is known to occur in a variety of species in neighbouring countries. To date, the genus *Lyssavirus* contains 18 species, and more will certainly be discovered soon.

These findings have changed our understanding of this virus group, but we still do not know much about rabies as a zoonosis. For instance, it appears that bats are the original reservoir for *Lyssavirus* species. It seems likely that RABV emerged long ago, following transmission from bats to flightless mammals. That said, all the more recently discovered virus species were isolated during laboratory research, not because they had provoked disease in humans. These species are found in Europe, Africa, Asia, and Australia. The only *Lyssavirus* species known to occur in the Americas, including in bats, is RABV, making for a unique situation. In any case, dog bites cause most, if not nearly all rabies cases in humans worldwide (resulting in 50,000–60,000 estimated deaths per year; the actual figures are not well known). Consequently, addressing rabies in humans requires controlling rabies in dogs, which means confronting the overly large stray dog populations found in many countries across the globe. Are such dogs true strays, or are they simply allowed to wander? Do they have owners or not? It is often difficult to answer these questions. However, we know for sure that most are neither vaccinated nor neutered and that rabies circulates among their ranks.

The diverse rabies viruses in bats do not represent a major public health risk because interactions between humans and bats are extremely rare. Bat biologists are vaccinated because their work involves handling these species.

PRION ZOONOSES

"Mad cow" disease, or bovine spongiform encephalopathy (BSE), is caused by a pathogenic prion (see p. 13). This infectious protein has an incubation period of several years. It causes localised lesions in the brain, for which no treatment is currently available. Infections are always fatal. The disease was first noticed in cattle in the UK in 1985. It remained in the headlines from the late 1990s to the early 2000s, when a link was finally established between bovines and human illness when researchers described variant Creutzfeldt-Jakob disease (vCJD). Initially, there was great uncertainty regarding how many humans had been infected. For instance, in 2000, Neil Ferguson's team at Imperial College London predicted future case numbers based on a variety of assumptions. It was suggested that, by 2020, between 63 and 136,000 cases of vCJD might occur in the UK population. This range was extremely broad because there was great uncertainty around many parameter values, including the disease's incubation period. INSERM conducted a study in 2001 in which a more precise estimate of incubation time was used: 17 years. The researchers then arrived at 205 predicted cases, which is not far off from the UK's current case count of 178. A few infections were also seen in individuals outside the UK: 28 in France and 26 in the rest of the world. In the UK, BSE also infected other animal species, including cats and zoo animals. In each case, the infection was traced back to contaminated food.

Historically, neurological illnesses associated with transmissible spongiform encephalopathies have been described several times: scrapie, a non-zoonotic animal disease, in the late 18th century; kuru, a human disease endemic to Papua New Guinea spread by ritual cannibalism, in the early 20th century; and, finally, classic CJD. It took some time for physicians and veterinarians to connect the dots between these different diseases. In the early 1980s, US neurologist Stanley Ben Prusiner laid the groundwork for understanding the role of prions and their conformational changes in disease aetiology. He was awarded the 1997 Nobel Prize in Medicine for his work.

Towards the end of 1985, UK veterinarians noticed unusual neurological symptoms in antelopes at the London Zoo as well as in bovines. These symptoms always ended in death. The following year, BSE was identified. Two years later, the relationship was established between the disease and the use of meat-bone meal in cattle feed. At first, the disease was linked to scrapie, a non-zoonotic prion disease that has never been shown to pass to humans.

BSE occurred very sporadically prior to the mid-1980s, which is when the large-scale distribution of animal meal began. Before the 1980s, meat-bone meal was sterilised at extremely high temperatures, destroying any prions. However, in 1980, operating procedures were changed to improve the industry's profitability, which left infectious prions in the meal. In 1991, the first case was observed in bovines in France. By 1993, more than 100,000 cases had occurred in the UK. Between 1986 and 2000, there were more than 190,000 known infections in bovines. Not included in this figure are the large number of animals that had been infected but that were slaughtered before any neurological symptoms could occur and before their quantities of prions reached detectable levels. If an animal tests negative, it means there is no risk of contamination. Even before systematic testing (e.g., before 2000), it was safe to eat the animals as long as any risky tissues were removed: the brain, spinal cord, thymus (sweetbread), and certain parts of the intestine.

The "mad cow crisis" was a wake-up call from public health, ethical, and economic perspectives. People became more aware of livestock feeding practices, and beef consumption plummeted. In the 2000s, drastic measures were taken to control the disease: all cattle in slaughterhouses were screened; the EU passed a complete ban on using meat-bone meal in animal feed; and any animals likely to transmit disease were systematically removed. These measures ended the epidemic.

There have since been only a few isolated cases of atypical BSE, which is distinct from classic BSE. These cases were identified thanks to massive screening efforts during the BSE epidemic. They occur regularly but at very low frequencies. Atypical BSE arises not from contaminated meat-bone meal, but rather from the natural course of aging. The prions spontaneously change conformations after a long incubation period via a similar mechanism to that seen in sporadic CJD in humans. The annual incidence of atypical BSE is about 1 case per million.

In Europe at present, BSE testing targets at-risk bovines (i.e., animals over 48 months of age that died on the farm or were euthanised due to disease or injury). However, the other measures remain in place. As a result, and given current epidemiological dynamics, the risk of vCJD in humans is now quite negligible.

PREVENTING AND LIVING WITH ZOONOSES

Zoonoses have existed since the dawn of humanity, and they are here to stay. It is up to us to determine how to live with them. To facilitate coexistence, we must naturally adopt certain individual behaviours. However, living with zoonoses will require, above all, a collective commitment to deploying the technical, technological, ecological, legal, and legislative tools at our disposal. The saying that an ounce of prevention is worth a pound of cure is certainly applicable in this context, whether prevention takes place on the individual or collective scale.

PREVENTION AND TREATMENT AT THE INDIVIDUAL SCALE

Limiting Transmission Risks

As individuals, we can protect ourselves from zoonoses via simple behaviours that involve basic hygiene and an elementary understanding of the ecosystems and animals with which we come into contact. Indeed, through our daily actions, we can limit our exposure to any zoonotic agents that may occur in our environment. These actions will differ depending on the mode of transmission: direct contact with animals, environmental exposure, ingestion of contaminated food or water, or arthropod vectors (see Figure 11).

Avoiding contact is the easiest way to prevent directly transmitted zoonotic diseases. For example, you should never touch a dead or injured animal with your bare hands, especially if it is a wild animal. This simple measure will protect you from physical injuries, such as those inflicted by bites, scratches, and beak or horn blows. In other cases, just avoiding any kind of touch is important. For example, hares and other species can be carriers of tularaemia, caused by the bacterium *Francisella tularensis*, which can penetrate bare skin. You should also avoid handling animals that are sick or behaving abnormally. One obvious example is that wild mammals infected with rabies act differently, which includes displaying less fear of humans. When interacting with pets, you should always adopt good basic hygiene, such as washing your hands after contact. In this way, you avoid introducing zoonotic agents (e.g., any helminth eggs on your pet's coat) into the mucous membranes of your mouth, nose, or eyes. You should also avoid letting your pets lick you, a behaviour that can infect small lesions on your skin or your mucous membranes with bacteria from your pet's oral microbiota. In work environments, direct transmission can be prevented by wearing personal protective equipment (PPE): specific clothing to don when dealing with animals; gloves for handling contaminated substances (e.g., dead animals); safety glasses to protect against splashes during pressure washing; and a filtering face mask when disposing of high-risk materials, such as abortion products. PPE can be adapted for use with pets, if necessary. Finally, you can reduce zoonotic risks by properly caring for the health of domestic animals, which includes rearing them under appropriate conditions and using targeted preventive and curative treatments when necessary.

You can also avoid consuming commonly contaminated foods, especially if you are in an at-risk category. Notably, people who are pregnant or immunocompromised should avoid eating raw milk cheeses, certain types of cold cuts, and seafood products, which often carry *Listeria* bacteria. These two categories of individuals are at risk of severe infections in a way that those with normally functioning immune systems are not. Another recommendation is to avoid collecting wild berries at low elevations, especially in the vicinity of travel corridors used by wild animals. Their droppings can transmit various parasites, such as *Echinococcus* species. For the same reason, you should carefully wash or peel any vegetables and fruits eaten raw, as they may have been contaminated by excrement. Boiling or cooking food eliminates most foodborne zoonotic agents. You should always cook meat extremely well, especially pork and poultry, to avoid consuming viruses (e.g., hepatitis E), bacteria (e.g., *Salmonella* or *Campylobacter*), and parasites (e.g., tapeworms or *Toxoplasma*). Take extra care when meat is barbecued, a technique that does not always cook foods through. Practicing good hygiene in the kitchen is also important. When refrigerators and work surfaces are improperly cleaned, it allows the growth of enteropathogenic microbes and can result in cross-contamination. Wooden cutting boards are a special concern because they are often used to cut meat but can be difficult to clean.

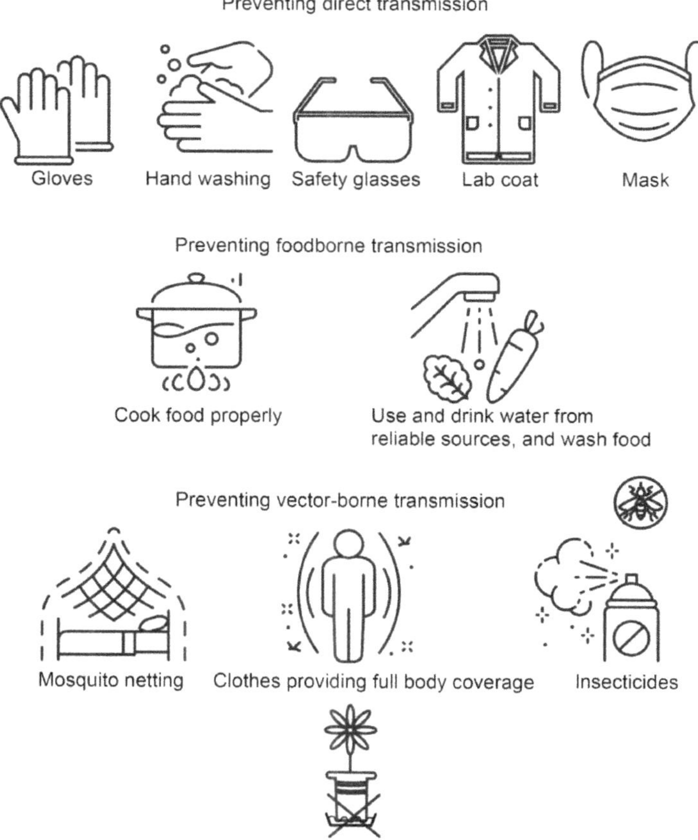

Figure 11. Individual actions for preventing infection with directly transmitted, foodborne, and vector-borne zoonoses.

Waterborne agents can be eliminated via modern water filtration and chemical treatment methods. However, you should never drink water from sources of unknown quality. When traveling internationally, avoid drinking untreated tap or well water. You should also steer clear of ice cubes, which are not always made with drinking water. Cold buffets spread out on crushed ice are another concern. Furthermore, it is preferable that you do not swallow any water used for showering or brushing your teeth. On such trips, adopt the same practices as when you are hiking and want to drink water from ponds or streams. Before drinking any water, you can treat it yourself via filtering, boiling, ultraviolet sterilisation, the use of disinfectant tablets, or a combination thereof. Such treatments can eliminate or inactivate any pathogenic microorganisms present.

For vector-borne zoonotic diseases, the best preventive strategy is to avoid infested areas, wear clothing that entirely covers your body, and apply repellents to your body and clothes. Certain mosquitoes can develop around human dwellings, so it is important to eliminate any sources of stagnant water, even small ones, because they serve as potential breeding grounds (e.g., drainage plates under flower pots, rainwater collection containers, and/or dog water bowls). In highly infested areas, bites can be prevented by covering windows and beds with mosquito nets, to which insecticides may also be applied. To avoid tick bites, you should wear clothing that entirely covers your body. In particular, you should pull your socks up over your lower pant legs. Because ticks remain attached for several days, you should inspect your entire body after each instance of potential exposure (e.g., walks in the forest or picnics at the edge of the woods). Performing a tick check will allow you to identify any that have attached themselves to you and to remove them as quickly as possible, thus limiting the time during which they can transmit viral or bacterial pathogens. The easiest approach is to use a tick removal tool such as a "tick twister": simply insert the tool's hooked end between the skin and the tick's rostrum, then twist the tool around several times. You can also use fine-tipped tweezers. Regardless of the technique, it is essential to remove any ticks quickly and completely.

Boosting the Body's Defences

Sometimes transmission cannot be avoided. Our bodies respond to the presence of infectious agents by mounting an immune response (see p. 22). However, some people are more sensitive than others to pathogens. Immune responses vary among individuals and within individuals at different points of their lives.

From the very beginning of our lives, our immune systems are learning to deal with the microbes they encounter at the interface created by our natural barriers, namely the skin and mucosa. Thus, each of us has a particular immunological history that is determined by our environments, including exposure to different animals, vegetables, foods, cognitive and emotional experiences, and, most of all, microbes. We are surrounded by a microbial landscape, composed of both pathogens and non-pathogens, that shapes our immune responses through cross-reactions to sets of similar antigens. This developmental process is greatly enhanced by the viruses, bacteria, fungi, protozoa, and acarians that live on us and make up the microbiota associated with our intestines, skin, respiratory systems, and other organs. These microbes also serve as a barrier against invasions.

In particular, the diverse antigens we encounter from our time *in utero* through our early childhoods seem to teach our immune systems to better tolerate intruders. These interactions influence our subsequent immunological ability to respond to the pathogens we encounter and have an impact on the degree of immune dysfunction. It is thought that allergic, autoimmune, and inflammatory diseases, and even certain cancers, are increasing in prevalence in Western countries because infants and children are exposed to lower levels of microbial diversity (i.e., the hygiene hypothesis). The world's last hunter-gatherers have the richest microbiota, while humans dwelling in large Western cities have the poorest. Thus, we might derive benefits from early exposure to microbial agents, including the potential pathogens transmitted by animals.

Once this period of immunological malleability has passed, our immune responses display dramatic variability and are influenced by many factors, including our disease history, age, gender, physiological status (e.g., pregnancy), and genetics (see sidebar p. 26).

Are there ways to bolster the immune system? There are many commercial products that claim to do so. However, scientists have frequently failed to find evidence that certain plants or drugs affect the quality of our immune responses. The immune system is complex, as are the factors that interact with it. Thus, it can be quite confusing to understand how "immune enhancement" could happen. One theoretical way to prevent disease could be to modulate the microbiota's composition, by administering microorganisms with physiological benefits (i.e., probiotics) or dietary fibres that promote the growth and development of specific microorganisms (i.e., prebiotics). Beneficial effects have been seen in people with certain diseases. However, it remains unclear at present how well these food supplements more broadly reinforce the intestinal flora of healthy people. Indeed, their effectiveness seems to depend on many factors, including microbial strain, preparation method, and individual physiology. It is crucial that their use be medically supervised.

As individuals, vaccination is the most effective strategy we can adopt to protect ourselves against zoonotic pathogens. It is mostly available for certain zoonotic diseases, generally those caused by viruses. Vaccination works by stimulating the body's immune defences without causing disease. The process involves injecting a small quantity of foreign matter, either from the target pathogen or a close relative, before the person has encountered the infectious agent in question. Consequently, the body adds the pathogen to its "memory banks", allowing a rapid specific protective response by the immune system in the case of future infection by the pathogen. A key step in vaccine development is identifying antigens that can serve as vaccine targets, a task that is easier for viral diseases than for bacterial diseases. It is extremely complex for parasitic diseases, which is why no vaccines against parasites have been developed to date.

Vaccines are medical treatments that must be prescribed by a physician, who will take into consideration a person's individual health concerns and risks. In Europe, vaccines against certain zoonoses are recommended for people whose work results in specific health risks and for travellers visiting places where transmission rates are high. For example, vaccination against rabies (a viral disease) is recommended for veterinarians, bat biologists, staff at animal shelters, and slaughterhouse workers. It is also a good idea for those travelling to countries whose dog populations have a high prevalence of rabies. Additionally, vaccination against leptospirosis (a bacterial disease) is recommended for people who regularly come in contact with potentially contaminated water during work or leisure activities. However, the antibodies elicited by vaccination do not protect against all the *Leptospira* serogroups, which is a major limitation of this vaccine. Vaccinating dogs helps protect their owners because infected dogs excrete leptospires in their urine.

UNDERSTANDING VACCINATION

The mechanism underpinning vaccination was discovered in 1796 by Edward Jenner, an English physician. He observed that people who were frequently in contact with cows were protected from human smallpox, a serious and often fatal disease. Such individuals had pustules on their hands that were caused by vaccinia (i.e., cowpox), a disease transmitted to them during the milking process. To test his hypothesis, Dr. Jenner inoculated an 8-year-old boy by scratching his skin and introducing pus taken from a milkmaid's arms. Three months later, Jenner inoculated the child with human smallpox; no signs of the disease ever appeared. This approach became known as vaccination (from *vacca*, the Latin word for cow), and its importance was eventually recognised by the scientific community. Nearly a century later, Louis Pasteur determined that vaccination could be based on inoculating individuals with "weakened viruses that cannot kill but rather that cause a mild form of disease that protects from the deadly form of disease". He successfully isolated, purified, and inactivated the rabies virus, which allowed him to develop the first rabies vaccine for humans in 1885.

For diseases transmitted among humans, the value of vaccination is largely rooted in population-level herd immunity (see p. 31), including to zoonotic pathogens such as SARS-CoV-2 or influenza A(H1N1)pdm09, which caused the 2009 flu pandemic. If a large percentage of a population becomes immune, disease transmission dynamics are disrupted, reducing the risk of infection for those who remain vulnerable. Using modelling, we can estimate the level of vaccination needed to prevent a disease from spreading, a figure that is disease specific but that climbs with pathogen transmissibility. Thus, the individual decision to become vaccinated is an action that we take for the collective good.

Medical Treatment

When transmission has occurred and resulted in illness, medical treatment may become necessary. Depending on the symptoms, physicians may or may not face difficulties in arriving at a diagnosis. As a patient, it is essential to provide details regarding the disease's onset and context to help determine whether it could be a zoonosis.

Once a disease has been identified, the doctor can potentially prescribe an etiological treatment to eliminate the pathogen: antibiotics for a bacterial infection, anthelmintics for worms, antifungals for a fungal infection, or antivirals for certain viral infections. A disadvantage of these treatments is that they can disrupt our microbiota because they affect not only the pathogen, but also other microbes of the same type. Physicians may request laboratory analyses to determine how sensitive the pathogen is and thus customise the treatment to avoid provoking resistance. At times, it is possible to use highly specific treatments. For example, a person bitten by a rabid dog will be given emergency care involving the injection of an anti-rabies serum containing specific antibodies (i.e., immunotherapy) with a view to blocking the virus from reaching the nervous system. The person will also be vaccinated after the fact, in the hopes that an immune response can develop faster than the virus can spread.

Other forms of treatment are used to ease the disease's symptoms (e.g., fever and/or pain). Some zoonoses require complex, long-term treatments. For example, alveolar echinococcosis may necessitate extensive surgery and/or prolonged chemotherapy.

PROMOTING VETERINARY PUBLIC HEALTH
AT THE COLLECTIVE SCALE

Veterinary public health (VPH) aims to predict and prevent the transmission of zoonoses to humans via the implementation of public policies at local to international scales. The French Veterinary Academy defines VPH as the suite of policies seeking to protect human, animal, and ecosystem health and well-being by taking collective action related to domestic and wild animals, including any animal products entering the food chain. VPH thus contributes to sustainable development and helps implement the One Health concept. This definition reflects the sentiment expressed in the WHO Constitution: "Health is a state of complete physical, mental, and social well-being and not merely the absence of disease or infirmity." VPH thus involves taking steps to a) better anticipate future zoonotic risks, b) protect the population from the latter, and c) communicate about any concerns.

Anticipating Risks Through Monitoring and Assessment

Disease surveillance programmes aim to collect reliable, real-time data to detect pathogens as early as possible, describe pathogen distributions in space and time, or verify pathogen presence or absence. More specifically, information on epidemiological indicators is systematically collected and analysed over spatial and temporal scales. These indicators may be focused on humans, domestic animals, wild animals, or environmental factors. They may estimate very different metrics: the number of deaths; the occurrence of certain non-specific syndromes (e.g., abortion or fever); cases reported by medical doctors or veterinarians; isolation of particular pathogen strains; the occurrence of genes related to virulence or antibiotic resistance; environmental factors associated with the presence of certain vectors; and incidents of food safety non-compliance. To strengthen surveillance efforts, it is especially important to develop participatory systems that directly involve different communities (e.g., everyday citizens, consumer groups, livestock associations, networks of practicing veterinarians, and professional solidarity funds). Such work should account for these stakeholders' specific concerns while firmly establishing collaborations rooted in the humanities, life sciences, and social sciences. These individuals act as boots on the ground because they are on the front lines of surveillance.

Surveillance efforts are event-based (i.e., passive) when they bring together pre-existing public health data and planned (i.e., active) when they carry out research and collect new data via targeted work. Examples of event-based surveillance in France include networks that record and analyse mortality data, such as the Observatory for Farm Animal Mortality (OMAR) or the Epidemiological Surveillance Network for Birds and Wild Terrestrial Mammals (SAGIR). The European equivalent of SAGIR is EWDA. Furthermore, there are official records of animal disease cases, which were reported as required by law. The main limitation of this approach is that it is relies solely on reported cases, which are not necessarily representative of actual cases in the field. Some planned surveillance efforts use sentinel animals (see sidebar p. 118) to determine pathogen occurrence or distribution.

SENTINEL ANIMALS

Sentinel animals are placed or chosen at a given location. Their status is then monitored over time for evidence of exposure to a particular pathogen; they thus act as a type of early warning system. Monitoring may involve determining whether pathogen-specific antibodies have appeared in the animal's blood between two sampling periods (i.e., seroconversion), whether the pathogen occurs in the animal's tissues, or whether the animal is clinically ill or has died. If a pathogen has never yet been detected, sentinel animals can sound the alarm by revealing potential introduction events. If a pathogen is known to be present, sentinel animals can be used to assess the degree of circulation. Several factors determine the sensitivity of sentinel-based surveillance systems, including monitoring interval, animal susceptibility to the target pathogen, animal number, and animal spatial distribution. Sentinel animals can reveal the presence of pathogens on the scale of an individual farm or an entire region. For example, in southern France, the seroconversion patterns of free-range poultry have helped detect the circulation of West Nile virus in mosquitoes, signalling the risk of infection for humans. In Asia, unvaccinated chickens are used as sentinels on chicken farms where the other animals are vaccinated against avian influenza A(H5N1). Their lack of immunity means they will die if the virus appears on the farm.

Zoonotic disease surveillance takes place at multiple scales. In France, surveillance in humans is identical for zoonotic and non-zoonotic diseases. Surveillance in animals is carried out by the National Animal Health Epidemiology Platform (ESA). Founded in 2011, ESA participates in international surveillance efforts. It also develops, adapts, and runs several surveillance tools aimed at various zoonotic and animal diseases. Also contributing to surveillance efforts is the National Food-Chain Health Surveillance Platform (SCA), which was created in 2018. At the global level, surveillance relies on close collaborations between various international organisations, notably the OIE, WHO, and FAO, which receive funding from the World Bank and the United Nations Development Programme (UNDP). In particular, countries must immediately contact the OIE if they observe any of the organisation's listed diseases. Furthermore, in 2006, the Global Early Warning System (GLEWS) was established to improve disease detection and jointly manage emerging risks at the ecosystem-human-animal interface. However, its visibility has declined since December 2018 because its website is no longer updated. As illustrated during the COVID-19 pandemic, a key component of surveillance is the availability of effective diagnostic and screening methods (see p. 26). Indeed, pathogens are more easily introduced and spread when infrastructures for monitoring public health and conducting laboratory diagnostics are lacking. Such resources are particularly crucial when dealing with emerging diseases, which require the rapid development and application of laboratory methodologies. For example, when SARS-CoV-1 emerged in 2002–2003, precious time was lost during the search for the causative agent because investigators first thought they were identifying a *Chlamydia* species and were then convinced that they were dealing with a new type of influenza (see p. 144). Part of OIE's work is to coordinate a worldwide network of accredited reference laboratories that diagnose major zoonotic and animal diseases.

Surveillance networks are essentially surveillance webs made possible by the contributions of many partners, each acting as a link in the monitoring chain. The latter is made up of multiple components: sample collection, laboratory analysis, data compilation and analysis, result synthesis, knowledge production, and information diffusion. Effective surveillance therefore requires complex infrastructure and multifarious human and technological resources. Long-term access to these resources can be a weakness in surveillance systems because it relies on consistent funding, the availability of qualified personnel, and stakeholder diligence in reporting information to health authorities. It is essential to shore up system resilience should economic or health crises arise, with a view to avoiding a snowball effect. The COVID-19 pandemic provides a clear example of how vulnerable surveillance systems can be. The crisis led to population lockdowns and the remobilisation of human and financial resources across the globe. Another example is the PREDICT surveillance and early warning programme. It was created in 2009 by the US Agency for International Development (USAID) and was terminated by the Trump administration in September 2019, just prior to the emergence of SARS-CoV-2. In addition, it is essential to be able to evaluate the technical effectiveness, public health utility, and social acceptability of surveillance systems if we wish to ensure their improvement.

To properly analyse zoonotic risks, high-quality surveillance is crucial. It yields information that will ultimately allow decision-makers to implement appropriate management strategies. Here, the term risk refers to the likelihood of an adverse event occurring, taking into account its deleterious consequences. In this case, the hazard is a threat to public health caused by one or more zoonotic agents. For a given hazard, risk is estimated using a four-step assessment approach: 1) the hazard's probability is calculated; 2) the probability of exposure to the hazard is determined; 3) the negative effects of the hazard are quantified; and 4) using qualitative or quantitative methodology, the hazard's probability of occurrence and harmful consequences are characterised for a given population. Depending on available data, modelling can be employed to various ends. For example, it can facilitate comparisons of different management scenarios. Risk assessment is a multidisciplinary scientific tool that draws upon published peer-reviewed research and other sources, notably expert opinions. The results should be interpreted in plain language that is accessible to all stakeholders. It is important to state all uncertainties and assumptions and to address how they may have influenced the final result. This transparency, alongside transparency regarding conflicts of interest, is an essential part of ensuring that the assessment is sound and that the recommended management strategies are consistent. In France, for instance, risk assessments related to animal health and zoonotic risks are performed by the National Agency for Food, Environmental and Occupational Health & Safety (ANSES), which submits its opinions and recommendations to the competent authorities as well as making them public. In the United Kingdom, such risk assessments are carried out by the Department for Environment, Food, and Rural Affairs (DEFRA). The EU equivalent is the European Food Safety Authority (EFSA). These agencies were created following several health crises (e.g., related to contaminated blood supplies, "mad cow" disease, foot and mouth disease, and dioxin). Their goal is to monitor and analyse risks independently of the governmental authorities tasked with risk management.

Disease Prevention and Protection

There are diverse strategies for preventing zoonotic disease transmission and protecting human communities because the sources of risks and modes of transmission are diverse themselves. The criteria that determine management options are feasibility, cost, effectiveness, social acceptability, and the minimisation of negative impacts (e.g., of an environmental, economic, social, and political nature). Ultimately, our interest in reducing zoonotic risks forces us to reflect on current and future animal production systems (see sidebar p. 127) and on our relationships with animals.

Here, the term prevention refers to preventing recognised zoonotic risks. It is distinct from the term precaution, which refers to measures taken to protect against hypothetical risks (e.g., if there is uncertainty regarding a disease's zoonotic origin). From a practical perspective, there is extensive overlap in techniques for preventing zoonoses and diseases strictly found in animals. They rely primarily on a combination of health measures and medical solutions, although social, economic, and political strategies may also be important. It is sometimes rather difficult to implement effective measures, especially in the case of zoonoses with many reservoir species or for which the mode of transmission is unclear. Human health, animal health, and ecosystem health are all linked. Theoretically, it is essential to favour approaches that integrate the medical sciences, the veterinary sciences, and ecology. However, there remains a long road between theory and practice (see sidebar p. 122).

FROM ONE HEALTH TO PLANETARY HEALTH

Research on zoonoses has highlighted the myriad interacting relationships among humans, public health, animals, animal health, and the environment. Several terms have been developed to summarise this network of links, each adopting a particular point of view. The One Health concept was first proposed in the 2010s and posited that physicians, veterinarians, and ecologists could work together to arrive at shared benefits. However, this perspective was far too anthropocentric, given that animal populations and environmental factors were largely seen as posing risks to public health. The Global Health approach incorporated the influences of globalisation but nonetheless focused exclusively on the benefits for human health. Finally, the Planetary Health approach brought in the social dimensions of health but did not yield clear recommendations. Instead of any of the above, we should espouse a view in which the intended recipients of any shared benefits are the planet, its living creatures, and its ecosystems. In 2021, the One Health High-Level Expert Panel (OHHLEP) defined One Health as follows: "[It] is an integrated, unifying approach that aims to sustainably balance and optimise the health of people, animals, and ecosystems. It recognises the health of humans, domestic and wild animals, plants, and the wider environment (including ecosystems) are closely linked and interdependent. The approach mobilises multiple sectors, disciplines, and communities at varying levels of society to work together to foster well-being and tackle threats to health and ecosystems, while addressing the collective need for clean water, energy, and air; safe and nutritious food; taking action on climate change; and contributing to sustainable development." This definition highlights the need to develop concrete policies that foster implementation.

.../...

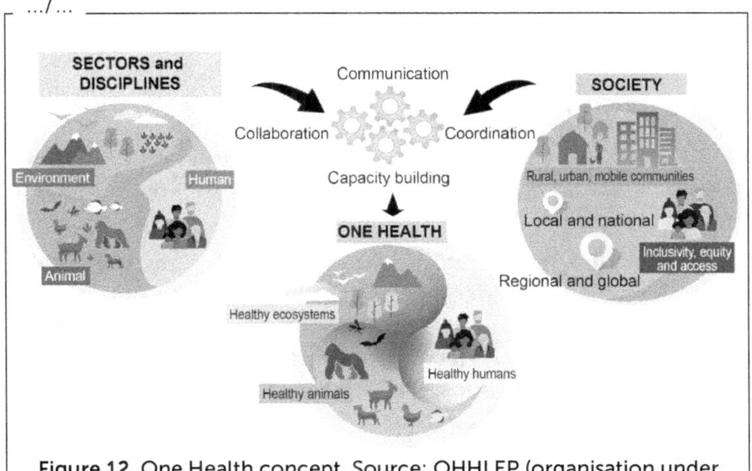

Figure 12. One Health concept. Source: OHHLEP (organisation under the joint aegis of the WHO, FAO, OIE, and UNEP).

Preventive Healthcare

As at the individual level, preventing zoonotic diseases at the community level is primarily based on good basic hygiene, especially when dealing with water- and foodborne pathogens. For instance, local governments must properly maintain shared public spaces and resources to limit contamination involving the faecal matter of domestic or wild animals, which harbour a variety of viruses, bacteria, and parasites. It is essential for populations to have access to high-quality drinking water (i.e., that meets microbial water quality standards). According to the WHO, 71% of the world's population had access to safe drinking water in 2017. All the points along water distribution systems are critical because they can be contaminated by the faeces of domestic and wild animals. In France, there is a regulatory framework for managing livestock effluent that limits the contamination of waterways and groundwater. For example, regulations stipulate that manure must be kept away from waterways and cannot be spread in rainy weather. In addition, composting-based treatments are encouraged. Treatment facilities have been established upstream of distribution systems and use physical processes such as agglutination, sedimentation, and filtration to eliminate the oocysts of zoonotic protozoa such as *Giardia* and *Cryptosporidium* (see p. 52). In addition, a combination of technologies are employed to disinfect the water, the main ones being ozonation, ultraviolet purification, and chlorination. Downstream, routine monitoring is carried out on the water distributed to households to confirm the absence of coliform bacteria, which are markers of faecal contamination.

As with foodborne zoonoses, prevention largely rests on applying basic hygiene practices in the case of structures, equipment, personnel, and products. Every professional along the food chain — including farmers, slaughterhouse workers, product processors, and vendors — has a responsibility to implement preventive measures to limit the risks of contamination. Indeed, in the EU, Hygiene Package regulations specify that food-sector professionals must put in place a customised food safety plan to ensure product compliance with health standards. In addition to respecting good hygiene practices, the food safety plan must be based on hazard analysis critical control point (HACCP) principles, which involve identifying the hazards (i.e., biological, chemical, and physical) associated with specific professional activities and implementing procedures for measuring and monitoring food safety. The entire food industry, from farm to fork, is tasked with following these food safety plans and must furnish tangible results. Thus, a traceability system needs to be put into place. The regulations specify that the system must allow identifiable entities to be traced, used, or located thanks to clear record keeping. Such records must be of sufficient quality that they can be employed for rapid and targeted product recall or withdrawal in the case of a food safety problem. Traceability is particularly important for processed foods, given that product transformation can promote the spread of microbes from a single infected animal to an entire batch of food products. Veterinary services may come in to verify that foodstuffs comply with regulatory requirements, particularly with criteria for microbiological safety. In addition, veterinary services may seize organs or other parts of carcasses that display abnormalities that could pose human health risks, such as the hazardous animal parts that must be eliminated to prevent BSE transmission (see p. 105).

Consequently, preventing infections in farm animals is a key strategy for limiting zoonotic risks, especially those associated with food products. Behaviours and habits must be adopted to reduce the risk of contamination; they generally involve the separation of animals and activities according to risk type. Implementation can take the form of biosafety rules, which describe all the measures employed to reduce the likelihood of pathogen introduction and spread in different contexts (e.g., farms, agrifood facilities, slaughterhouses, veterinary clinics, or laboratories), regions, and food chains.

Biosecurity is based on five principles:

- Exclusion (external biosecurity): preventing pathogens from reaching facilities by taking such measures as quarantining or vaccinating newly arrived animals, verifying food quality, averting the potential for contact with wild animals, and wearing appropriate clothing.
- Compartmentalisation (internal biosecurity): preventing pathogen circulation by compartmentalising facilities, where specific areas are dedicated to events associated with greater health hazards (e.g., quarantine area or birthing area); movement patterns among these areas flow from least to greatest risk.
- Containment: preventing pathogen diffusion away from facilities, which involves cleaning and possibly disinfecting any equipment that leaves the premises, managing waste and effluent, monitoring animal departures, and potentially installing, as seen at certain locations, internal vacuum systems that direct air flow from outside to inside
- Protection: preventing pathogen transfer to humans, which essentially involves practicing good hygiene.
- Preservation: preventing the environmental persistence of pathogens by carefully managing waste and effluent and limiting contact with wildlife.

Obviously, biosecurity implementation must be customised to deal with the specificities of different locations. In particular, laboratories dealing with microbes that are potentially dangerous for humans, animals, or the environment must take certain precautions to protect their staff and their facilities. For example, to reduce risks, some laboratories have been set up on islands or in fairly isolated locations, outside of large urban areas (when the focus is human diseases) or outside of livestock farming zones (when the focus is animal diseases). In certain cases, negative pressure laboratories have been built: the pressure differential between the interior and exterior of the buildings prevents any air from the escaping to the outside. All fluids and waste are fully disinfected prior to disposal. Staff are sometimes required to wear full-body, air-supplied suits to avoid any risk of infection. When an accident is seen, it is because these protocols were not properly respected.

On livestock farms, manure and slurry are major sources of environmental contamination. They must be appropriately dealt with. Prolonged storage or composting leads to fermentation, which is accompanied by increased temperatures. These thermal conditions are likely sufficient to destroy most zoonotic agents, However, further research is needed to clarify the effectiveness of these treatments. In addition, it is also an essential health measure to remove dead animals and infectious materials, such as abortion or birth products.

Sometimes, governments may impose certain measures. For example, to control avian influenza, most countries require those rearing birds to implement biosecurity measures. The specific measures to be taken vary depending on bird abundance (commercial breeding operations vs. backyard poultry), bird type (poultry, game, or zoo) and facility type (commercial, backyard, enclosed building, outdoor enclosure, or aviary). In addition, governments may require poultry and pig farmers to obtain biosecurity training, notably when animals are being reared under industrial conditions. If it is suspected that poultry have been infected with highly pathogenic avian influenza, investigations will be performed that target all types of rearing operations, and adapted measures will be taken in accordance with FAO and OIE recommendations.

However, this hygiene-based approach essentially focuses on health monitoring, documented outcomes, and disinfection efforts after pathogens are detected. Unfortunately, it is not suitable for certain situations, such as dairy farms producing raw milk cheeses. Indeed, cheese processing requires a level of microbial richness that is at odds with microbe elimination. To safeguard the gastronomic heritage that is French cheeses, compromises must be made to preserve the microbial biodiversity underlying cheese production. Fortunately, other forms of animal farming are being (re-)developed, systems that are more respectful of ecosystem functioning and animals' physiological needs. We are currently rethinking health risks and biosecurity practices.

A ROLE FOR AGROECOLOGY

The industrialisation of livestock farming has resulted in high-density, extremely specialised production systems that occur at various scales, from farms to regions. In these systems, animals of the same species are packed together and represent genetically homogenous breeds created via intensive artificial selection. Production levels are extremely high. Finally, animals are reared indoors, an approach that requires massive input levels. Industrial livestock farming thus creates conditions favourable to the spread of pathogens. Drastic biosecurity measures must thus be implemented to minimise the risk of pathogen introduction.

In response, many initiatives are seeking to implement agroecological principles to strike a balance between ensuring production and protecting biodiversity. The broader objective is to promote natural biological regulation (see p. 151) and foster farming conditions that better respect animals' physiological needs. This agroecological transition reflects an improved scientific, and namely ecological, understanding of how infectious diseases are regulated. It harkens back to more traditional forms of animal farming and is tackling different types and ranges of health risks. First, it is utilising animals' natural defences, by rearing them under conditions that do not run roughshod over their physiological needs. For instance, animals are allowed to produce lower yields, and farmers are increasingly raising traditional and local breeds. Second, there has been a reduction in both animal densities on farms and farm densities within regions, which limits the spread of infectious diseases. Third, it is possible to promote diversity-mediated functional services by increasing diversity at all scales—from intraspecific genetic diversity to interspecific diversity within natural plant and animal communities. Such can be seen with free-range and mixed-species farms. This transition represents a momentous change. It requires accounting for different risks, such as increased pathogen exchange among species, and characterising interactions between pathogens and ecosystems. Ultimately, agroecology aims to manage ecosystem health by more holistically addressing animal health, a process that can also end up revealing societal and political concerns.

In our globalised world, animals and products are constantly being traded. Consequently, we face the omnipresent risk of introducing pathogens into new habitats. The health and safety of all countries is thus intimately tied together, as underscored by the COVID-19 pandemic. From a collective perspective, it is crucial to recognise that commercial health and safety regulations are a key part of preventing animal and zoonotic diseases. The OIE is responsible for establishing health and safety standards that ensure health risks are limited during global exchanges of animals and animal products. Animal identification and traceability are essential tools in this work as they promote food safety and animal health (including in relation to zoonoses). All countries have limited human, technical, and financial resources. Thus, public policies prioritise certain diseases, for the most part zoonoses, based on various criteria, including current epidemiological circumstances. In the EU, infectious diseases in animals are assigned to one of five categories depending on their pathogenicity, their zoonotic potential, and their associated prevention and control measures. For instance, exotic diseases are subject to mandatory surveillance and reporting. Eradication programmes have been established for brucellosis, tuberculosis, and rabies. Methods for applying these measures are described in legislative and regulatory texts. Veterinary services are tasked with their enforcement, as part of animal health requirements. In the case of an outbreak of an exotic disease or a disease subject to mandatory reporting, the government will impose restrictions to prevent the pathogen from spreading and clear farms of health threats (see sidebar p. 129). Farmers face sanctions if they do not respect these rules. While the government financially compensates farmers, the funds received never fully cover the losses incurred. Furthermore, a herd's value is never just monetary; it is also emotional, psychological, and genetic. That said, these measures have functioned quite well from a public health perspective. They have led to a pronounced decrease in major zoonoses transmitted by domestic ruminants, such as brucellosis and tuberculosis.

ANIMAL HEALTH REQUIREMENTS

When certain diseases known to have major public health or economic impacts are detected on farms, governments may implement well-defined, highly restrictive measures. Affected farms are treated as outbreaks around which the government defines protection zones (i.e., the movement of animals is banned) and surveillance zones (i.e., the movement of animals is restricted). These zones span several kilometres. On the affected farms, any livestock present are counted and may all be slaughtered. Depending on the situation, this process takes place in a slaughterhouse, in a rendering plant, or on site to avoid any risk of pathogen spread. The bodies are destroyed. The farms are then disinfected and left empty for a "cleanout period" before they are repopulated with new animals.

Preventive Medicine

For some zoonoses, preventive healthcare is inappropriate or insufficient for stopping pathogen transmission to humans. Such cases may require the use of preventive medicine, including vaccines or medications. Above, we discussed using vaccination in humans to prevent the occurrence of certain zoonoses (see p. 114). Vaccination can also be deployed to establish herd immunity in the animal populations responsible for transmission to humans, thus severing the chain of infection. This approach is used with rabies, which largely infects humans as a result of dog bites. The WHO has found that, in countries with a high prevalence of rabies, vaccinating dogs is the most efficacious and cost-effective strategy for preventing human infections. Not only does it reduce human deaths due to dog rabies, but it also diminishes the need for post-exposure treatments in the case of dog bites (see p. 103). In 2015, the WHO, OIE, and FAO held a global conference during which an international consensus was reached: 2030 was set as the target year for eliminating dog-transmitted rabies cases in humans. This goal seems feasible, even if the precise number of human rabies cases remains difficult to estimate. That said, there are challenges related to the long-term implementation and funding of the global strategic plan, which serves to illustrate the complexity of collaborative initiatives on animal and human health (see sidebar p. 122).

Similarly, human populations are protected when ruminants are systematically vaccinated against Rift Valley fever virus, which causes a zoonosis that occurred in Mayotte in 2018–2019 and that continues to crop up in Africa. Vaccinated animals no longer serve as amplifying hosts when bitten by infected mosquitoes. Human smallpox was declared eradicated in 1980 following vaccination campaigns conducted by the WHO. Similarly, the OIE announced in 2011 that rinderpest, a non-zoonotic viral disease of artiodactyls, was eradicated. It is clear from such health victories that the systematic vaccination of domestic species can bear fruit and is often better accepted by local populations than are human vaccination campaigns.

Furthermore, some degree of human protection can be afforded by treating domestic animal populations that are infected with zoonotic pathogens, as mentioned in the section about disease prevention at the individual scale. However, such strategies must be employed with caution because the broad-scale use of antibiotics, anthelmintics, and insecticides can have significant environmental consequences and promote the emergence of resistance (see p. 82).

Prevention—the Environment and Wildlife

Broad-scale actions for preventing vector-borne zoonoses mainly take the form of disinsectisation programmes, during which the egg, larval, or adult stages of arthropod vectors are eliminated from the environment. Unfortunately, spray programmes have adverse effects on non-target arthropods, and the chemicals' effects can become magnified in primarily insectivorous animals, such as certain bird and bat species. Equivalent concerns exist even for biological control strategies, such as the use of *Bacillus thuringiensis* (Bti) spores to deal with mosquitoes. Tick populations can only be controlled by holistically managing ecosystem-level biodiversity. This integrated management approach is also useful in the case of other vectors.

In addition, wild species are often unnoticed reservoirs of zoonotic diseases that pass to humans, with farm animals sometimes acting as conduits. When zoonotic pathogens are harboured in wildlife, it can be more challenging to implement government-mandated control measures in farm animals. It may thus become necessary to apply strategies that promote or medically manage the health of wildlife populations. Many European countries have wildlife health surveillance programmes. However, the latter vary in scope and not all are coordinated at the national level. Consequently, in Europe, information about the health status of wildlife populations remains limited. Management measures targeting wildlife tend to be costly and poorly regulated. They must also navigate sociological complexities because different stakeholders (e.g., hunters, farmers, and naturalists) have different relationships with wildlife. Indeed, it is important to engage with stakeholders because the above broad-scale preventive measures often rely on labour volunteered by these groups of individuals. Failure to do so can result in counterproductive stalemates. In addition, certain wild species may be subject to highly different management regimes, depending on the context (e.g., species that are commercially hunted).

To eliminate zoonotic pathogens, it is largely impossible to completely eradicate all the members of a wild reservoir species, in contrast to what is done with livestock on farms. First of all, such an approach would be ethically questionable and socially unacceptable. Second, it represents a major conservation issue because destroying an animal population also means losing the genetic wealth it represents. Third, it would be challenging to carry out and would likely only be effective in the case of small, well-defined, and easily accessible populations of a known size. In France in 2006, this management strategy was applied to red deer (*Cervus elaphus*), a species that serves as a reservoir for bovine tuberculosis. More particularly, efforts targeted a population found in Normandy's Brotonne forest, which was being managed for hunting purposes and was isolated by physical barriers (i.e., a river and a highway). However, even in this case, the population's size was more than double the original estimate, which changed how the outbreak had to be handled.

Setting aside the example of population eradication, managing population densities can be a useful approach. Theoretically, it is possible to prevent a disease's persistence and spread within wildlife by forcing the disease's R_e below a certain threshold. However, this metric can be complex to estimate (see p. 31).

Under real-life circumstances, the culling of wildlife populations often leads to increased pathogen persistence or spread. Indeed, such efforts can disrupt ecosystem dynamics, resulting in cascading responses in animal behaviour, social structure, territoriality, migration, or reproduction. The impacts extend beyond the target species to all the other species, large and small, with which it interacts. There may be major consequences for biodiversity, and the situation can give rise to various health and conservation issues. In addition, such strategies are costly and difficult to maintain over the long term. In Europe, counterproductive effects were seen in culling campaigns focused on red foxes (*V. vulpes*), aiming to control alveolar echinococcosis; an Alpine ibex (*Capra ibex*) population affected by brucellosis; and badgers (*Meles meles*) occurring in proximity to bovine tuberculosis outbreaks, to name a few examples. An alternative to eliminating the entire population is to solely eliminate infected individuals. However, this approach is complicated to implement, given that it requires the use of an efficient screening test (see p. 26) that is adapted to the target species and field conditions. The animals must also be captured, which can be quite expensive, given the technical and human resources required. For example, on average, it costs around €720 to serologically screen an ibex for brucellosis, as the animal must be trapped first. This figure is €4.60 for a domestic ruminant.

Control campaigns often target rodents because they host various microorganisms and tick larvae, which themselves vector pathogens. Anticoagulants are the chemical compounds that tend to be deployed. Unfortunately, they can also end up killing the rodents' predators, which undermines control efforts. In urban areas, the targets are rats and mice. However, control efforts can only be effective if human food waste is also limited, and potential refuges are addressed. In rural areas, the main rodents of concern are voles (genera *Microtus* and *Arvicola*).

Their high densities largely result from intensive agricultural activity, including landscape simplification, hedgerow removal, and the elimination of predators.

Vaccination can be a helpful management strategy for changing transmission dynamics when a disease is established within wildlife, especially when populations exist along a continuum. In the latter situation, density reduction becomes an ineffective strategy. However, mass vaccination is costly and complex to implement. First, an effective vaccine against the disease must be available and suitable for use in the target wildlife species; oral administration is often a prerequisite. Second, ideally, the vaccine must be safe for both the target species and non-target species because various animals may end up ingesting it. Meeting this requirement is no easy task. Third, orally administering the vaccine means designing baits that will appeal to the target species and that can be distributed over a very precise grid within the target zone, either on foot or by plane. Given the cost in time and resources, it is difficult to maintain immunisation efforts over the long term. Thus, this strategy should only be prioritised if it is possible to attain a level of vaccination coverage that achieves herd immunity within the population over the short term. In 2001 in France, this approach was successfully used to eradicate vulpine rabies (see sidebar p. 134). At present, research is exploring the possible use of an injectable or oral vaccine to protect badgers against bovine tuberculosis.

Sometimes, other medical strategies are utilised to disrupt the transmission dynamics of zoonotic diseases in wildlife. Administering drug-based treatments raises ethical and environmental concerns as this approach involves disseminating pharmaceutical compounds into nature. There could be negative effects on the biology of target and non-target species alike. It could also lead to the appearance of resistance (see p. 82). In addition, as in the case of vaccination, this method is costly and complex to implement. It also requires the capture and release of many animals or the distribution of numerous drug-containing baits. This approach is sometimes used in efforts to control alveolar echinococcosis. Foxes are given an anthelmintic, praziquantel. Although parasite prevalence greatly declines, the worms are not eradicated. Although flatworms rarely seem to develop resistance to pesticides, it is hard to predict the potential impacts of the compound's broad-scale dissemination. Another medical strategy under debate is administering immunocontraceptives to dampen reservoir host reproduction and thus affect population dynamics. In the UK, this method has been proposed as a way to address the role played by badgers in bovine tuberculosis epidemiology, given societal disapproval of badger culling. However, preliminary research has suggested that the feasibility and efficacy of this approach relies on females being treated at least every two years. This method is not without its detractors, as the immunocontraceptives could potentially affect non-target species or end up in the environment, which is already filled with endocrine disruptors.

ERADICATION OF RED FOX RABIES IN FRANCE

What a nasty surprise when rabies returned to Pasteur's home country in 1968. It arrived in the form of a fox variant. At first, control efforts focused exclusively on killing, gassing, poisoning, and trapping foxes, without accomplishing any discernible results. Gradually, the pioneering work of Swiss veterinary teams gave steam to the idea of vaccinating the foxes. First, a new vaccine had to be developed. It needed to be effective via oral administration and remain viable under external environmental conditions, given the absence of the more traditional cold chain. Then, it was necessary to develop an appealing bait to which the vaccine could be added. In the field, the baits had to be placed in such a way as to reach most of the foxes found in a given area. The required tools were developed in the 1990s. Initially, the bait was distributed on foot. Later, it was dropped via helicopter. Within two years, the disease had disappeared. The last case was recorded in 1998, thirty years after the reappearance of rabies. In 2001, France was declared officially free of the disease. To avoid any risk of re-emergence, vaccination campaigns were carried out until 2003.

For game species, additional actions can be taken to reduce zoonotic risks. For example, in the case of bovine tuberculosis, preventive actions consist of collecting animal viscera following a hunt, such that they can be destroyed at a rendering plant, and prohibiting the supplemental feeding of wild animals (a hunting practice) near sites of disease outbreaks. However, it is possible that the opposite effect could be achieved: if supplemental feeding stops, animals from the zone where the disease is present may move to areas where feeding is maintained, thus introducing the pathogen. In theory, it would be possible to put up fences and delineate high-risk zones. However, such a strategy is complex to apply and maintain over the long term; it might also be difficult for the population to accept.

All issues considered, the most appropriate or most feasible approach is often to just leave wildlife alone and to focus instead on farm biosecurity, including efforts to limit contact between domestic and wild animals. To this end, farmers can install and maintain solid perimeter fencing, monitor self-feeders in pastures, remove salt stones from mountain pastures, block wildlife from accessing livestock watering points, and keep free-range poultry in confined areas. However, it is impossible to guarantee that no contact at all will occur. Furthermore, these measures must be implemented by farmers, who may not agree with them or have the resources required for implementation.

Another essential facet of these efforts is to effectively deal with illegal wildlife trafficking. The possession and transport of wild animals is strictly regulated (see sidebar p. 157).

Communication

In any risk analysis, the last step is typically communicating the results. The first two steps are hazard identification and risk assessment, both performed by experts. The third step is risk management, which is carried out by decision-makers. However, all four steps must sometimes take place simultaneously, notably in times of crisis, as people face new diseases associated with many unknowns. It is particularly important to communicate with the general public about health risks so that individual preventive measures can be taken in response to collective-level strategies. Individual behavioural choices are particularly important for avoiding infections with tick-borne zoonoses (see p. 109). The same is true for foodborne zoonoses. Infections frequently

occur as a result of household conditions, such as poor food storage, improper cooking practices, or cross-contamination between foods. There is another set of essential tools for reducing zoonotic risks: boosting the awareness, knowledge, and training of farmers and other professionals who work with animals or animal products across a variety of industries.

Health education can take the form of governmental programmes that encourage behavioural changes aimed at reducing exposure risks or that limit the consequences of exposure at the population level. These programmes can take various forms, including communication campaigns aimed at the general public or the most vulnerable members of the public; expert advice provided by health professionals; continuing education; extracurricular courses; and food labelling. Although a range of programmes may be used, their effectiveness may remain limited. Indeed, the top-down transmission of information, from "experts" to "laypeople", is often doomed to failure because individuals differ in how they gauge risks and relate best to outreach that is rooted in their own life experiences. This challenge becomes especially clear when communication involves "invisible threats", like pathogens. For example, even within the community of bat researchers, perspectives on the taxon's zoonotic risks vary greatly, depending on the specific field of study. Unidirectional messaging will only be effective when people are open to receiving it, namely because they were already receptive. That said, people can be "nudged" towards specific behaviours that promote health by fostering certain conditions. For instance, making meat thermometers available can encourage people to assess whether their meat is properly cooked. Alternatively, positioning sinks in a more accessible way can encourage people to wash their hands before they enter the lunchroom.

Historically, researchers have tended to communicate primarily with their peers. However, sharing research with the public is now of paramount importance. Furthermore, given the current paucity of financial resources for research, highlighting the relevance of one's work to potential funding organisations has become almost vital for scientists. Therefore, communication by researchers is not necessarily altruistic. Indeed, it is always easier to find funding for scientific topics that have received media coverage and that have an apparent societal impact. To illustrate, prion research would not have received the same levels of funding without the media focus on the "mad cow crisis".

However, communicating about health risks is a particularly delicate task, especially in crisis situations. When the term "health crisis" is used following an event such as the emergence of a new disease, it is because the government has failed to nurture a sense of security and trust within the population. The notion of a "crisis" evokes images of political and social destabilisation in urgent need of a response and carries weighty significance within the context of the media. Crises generally arise in situations characterised by uncertainty and result from differences in how everyday citizens and public-sector experts perceive existing risks. The latter are scientists who have been tasked with sharing their collective expertise with decision-makers and thus informing public policy. This work involves making effective use of all available scientific data and knowledge while remaining fully transparent with regards to any uncertainties. One of the late 20th century's major health crises arose in relation to "mad cow" disease. In this case, the actual threat ended up falling far short of the dire predictions made (see p. 105). A similar scenario occurred in 2009, when the A(H1N1)pdm09 virus emerged, and the WHO declared an influenza pandemic (see p. 96). Some countries feared that the strain would be highly pathogenic and overreacted, given that the strain exhibited low virulence in humans. Such cases illustrate how challenging it can be to plan for the pathways, impacts, and real-time management of emerging infectious diseases.

During health crises, public authorities intend to reassure the population through their actions but often end up having the opposite impact. Such situations are examples of a "security paradox": the more those in power seek to engender a sense of security, the more they actually generate insecurity, especially if the underlying issues are not clearly identified and explained. If the government is transparent about existing gaps in knowledge, the population retains the message that the authorities are not well informed, creating insecurity. However, if the government fails to transparently communicate about sources of uncertainty, the population feels as though it is not receiving the full story and ends up suspicious, a situation that feeds conspiracy theories. It is therefore essential for governments to remain transparent in their announcements and decisions — clearly expressing what is and is not known. Unfortunately, urgent communication is often overly rushed and insufficiently planned, which sometimes results in ambiguous or contradictory statements. As a consequence, institutions are further discredited in the eyes of the public, and distrust of public authorities deepens. For example, during the 2006 A[H5N1] "bird flu" crisis, a French government official recommended that meat be cooked thoroughly, a statement that probably contributed to the subsequent drop in poultry meat sales. However, the virus is airborne, not foodborne. Indeed, the phrasing was particularly clumsy: "there is nothing to fear, especially if the meat is well cooked". Cooking poultry meat properly is important, but only because it eliminates other zoonotic agents, such as salmonella.

Unfortunately, media coverage of scientific results has become just another commodity. This situation is at odds with the need to deliver accurate information to the public and to ensure transparency regarding uncertainties. In fact, certain media outlets exist solely to profit off of their large audiences and thus seek to garner a maximum of attention. To this end, they tend to utilise a narrative style in which information and figures are sequentially provided with the aim of surprising, shocking, or frightening the public. They may also pass the microphone to self-proclaimed experts who deliberately seed conflict. Professor Osterhaus at the Erasmus MC Research Centre in the Netherlands teaches a workshop in which young researchers learn to prepare for a 10-minute interview with a journalist. He recommends coming up with a single sentence that conveys the key scientific message, which should be repeated over and over for the full 10 minutes. In this way, the final message cannot be edited out given that it is the same as the initial message. Indeed, comments taken out of context can easily be repeated and turn into a "truth" that reappears over and over in certain mainstream media. This outcome has become all the more likely given that some journalists obtain information (e.g., sound bites and quotes) directly from specialised new agencies. As the information's context and source are not always verified, it is easy for errors and misunderstandings to occur. For example, several media outlets wrote headlining articles about the link between declining vulture populations and increasing human rabies cases in India. However, the journalists were treating the correlation between the two factors as fact even though no relationship had been established. Another serious problem is that certain journalists establish an equivalence between scientific findings and arguments arising from scientism and transhumanism. Indeed, it is irresponsible to spread the belief that humanity will find solutions to past and present ecological disasters and that, as a result, it is unnecessary to rethink our ways of living and the paradigm of unlimited growth upon which they are based. On the contrary, we must place practical and ethical limits on technological development. We need to take the time to fully consider the major challenges represented by climate change and biodiversity collapse.

While modern technologies facilitate access to information and allow its widespread dissemination, they also hinder higher-quality communication, which involves exchange, dialogue, respectful debate, and constructive criticism. Governance in public health is only efficient when it draws on expertise from multiple sources. It must also build reciprocal exchanges and a relationship of trust among scientific experts, everyday citizens, important third parties (e.g., non-profit organisations or labour unions), administrative bodies, and decision-makers. Guided by the humanities and social sciences, work is underway to develop these collaborative approaches to defining public policies (e.g., living laboratories). However, there is still a long way to go before these approaches are fully integrated into the policymaking status quo.

LIMITING ZOONOSIS EMERGENCE: A COLLECTIVE GLOBAL RESPONSIBILITY

The mainstream media is finally turning its full attention to the subject of emerging zoonoses and their ties to human activities. This shift occurred after the SARS-CoV, MERS, Zika, Ebola, and influenza crises and despite clear advances in prevention and treatment.

LESSONS FROM THE ANTHROPOCENE

Homo sapiens appeared 300,000 years ago and long lived in populations of a few thousand individuals. However, the species' impact on the planet dramatically increased with the first Agricultural Revolution, which occurred 12,000 years ago. It continued to grow as the first agrarian civilisations developed and then accelerated as a result of colonialism and early globalisation. Around 1800, the human population reached 1 billion. It is expected to reach 8 billion by 2024. Largely traveling by foot, *H. sapiens* took tens of thousands of years to spread across the world, leaving Africa to later arrive in the Americas and Australia. As of 2019, more than 4 billion humans have travelled by plane and can traverse the planet in a matter of hours. For two centuries, our species has radically modified all the Earth's ecosystems with ever-increasing speed and intensity.

Indeed, humans have shaped natural systems to ensure their own safety, security, and personal comfort. To this end, they have forged such tools that they are now the main agent of change, surpassing other geophysical forces. As a result, we seem to have entered a new geological era after just a few decades: the Anthropocene. The idea of a new geological era was proposed in 2000 by Nobel Prize-winning chemist Paul Crutzen, who posited that humans have become so numerous and active that they now rival the major forces of nature in terms of impacts on the Earth's functioning. Indeed, starting with the Industrial Revolution, anthropogenic activities have left a recognisable signature in the planet's rock layers. Traces can be seen even in the ice cores of Antarctica. Thus, within a very short geological time span, humans have disrupted the Earth's ecosystems in ways that will persist for tens of thousands of years. We are facing unprecedented planetary disorder as a result of massive deforestation, the excessive damming of rivers, and the pollution of the atmosphere, water, and soil. As a consequence, large numbers of animal and plant species are going extinct, endangering the resilience of all types of natural systems. In the Anthropocene, the Earth displays unpredictable functional responses to the disturbances created by a segment of humanity, and we are fast approaching the tipping points for climate change and ecosystem collapse.

This expansion has been spectacular and unbalanced, as well as sometimes imposed and poorly controlled. It has led to great pressure on the environment and other animal species. As a result, there have also been effects on the patterns and dynamics of zoonotic transmission, and the likelihood has increased that local transmission will become global.

DEFINING ZOONOSIS EMERGENCE

The term "emergence" refers to the appearance of an infectious agent: it may either be entirely new, or it may be known but increasing in a way that is unexpected, atypical, or fast. These shifts can manifest themselves in geographical distributions, clinical characteristics, or responses to established treatments. Concern over the emergence of zoonotic diseases centres on both the increasing frequency of zoonosis epidemics (i.e., abundance) and the increasing number of zoonoses (i.e., diversity). Here, we used the same definition of an epidemic, or an outbreak, as the WHO: "the occurrence of cases of disease in excess of what would normally be expected in a defined community, geographical area, or season". Note that this definition makes no reference to an established number of cases but does mention a pre-existing chain of transmission.

Public health systems come under substantial pressure when the abundance and diversity of zoonoses rise. Not only must they deal with the known challenges of existing zoonoses, but they must also navigate the unknowns that are part and parcel of emergent diseases. Fortunately, scientific advances, notably in molecular biology, have made it possible to faster identify and better characterise pathogens. Indeed, we now have access to more detailed descriptions of diverse potential pathogens (Ebola and Marburg viruses, bat *Lyssaviruses*, *Borrelia* — causative agent of Lyme disease). Nonetheless, it remains difficult to fully understand transmission cycles and the factors underlying emergence. It is complicated to arrive at generalised ideas given the intricate, multicausal nature of zoonosis emergence. That said, since 2000, numerous research findings have allowed us to identify the major features of emergent infectious diseases in general and zoonoses in particular.

INTERFACES

Pathogens can be found among microorganisms and parasites, which are contributors to biodiversity on Earth. The presence of living creatures, and especially vertebrates, entails the simultaneous presence of microorganisms, including some that are potentially pathogenic. Thus, any factors that affect these sources of biodiversity can influence the dynamics of zoonoses.

Let us examine an example of a zoonosis responsible for a pandemic. The underlying process can be broken down into three conceptual stages (see Figure 13). In the first stage, a potential pathogen moves from an animal to a human (i.e., a spillover event) during an encounter, which may be mediated by a vector or environmental conditions. It is important to understand the nature of this interface, including the major factors at play, if we wish to identify the actions that can help prevent emergence events. The interface can be separated into three intersecting components: 1) the hazard, otherwise known as the pathogen; 2) the encounter, or the contact between the pathogen and humans; and 3) human susceptibility to the pathogen. In the second stage, the zoonosis is amplified within the human population if there is human-to-human transmission. The likelihood of the latter will depend on the pathogen's ability to adapt to humans, the population's characteristics (e.g., density and/or mean state of health), and the health management regime in place. In the third stage, the above epidemic will become a pandemic once the pathogen has spread to several continents via the movements of animals or humans.

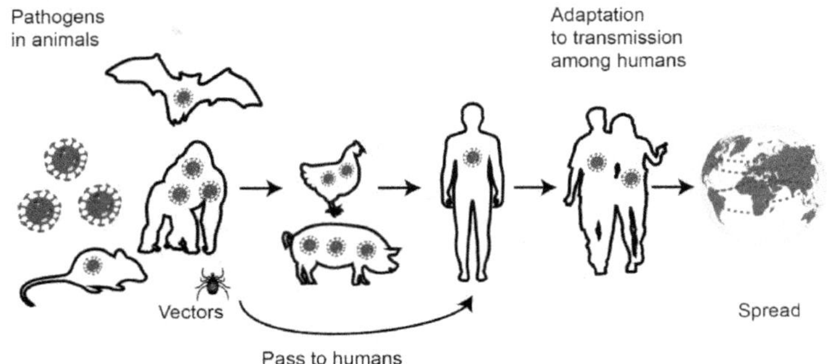

Figure 13. Schematic illustrating how a zoonosis can emerge and cause a pandemic in humans.

DETECTING NEW ZOONOSES

Identifying new diseases

When a completely new disease emerges, a certain number of cases must be observed before an illness is recognised as "abnormal" and the alarm is sounded (see p. 117). For example, hantaviruses were first discovered during the Korean War (1950–1953), when more than 3,000 members of the United Nations forces came down with a haemorrhagic fever with renal syndrome whose origin was unknown. It was only 25 years later, in 1978, that the infectious agent was discovered, Hantaan virus. It was named after the city in which the first cases were described. Researchers also identified and described the virus' reservoir, the striped field mouse (*Apodemus agrarius*). Similarly, it was clusters of "abnormal" illnesses that led to the discovery of Hendra virus in 1994 in Australia; Nipah virus in 1998 in Malaysia; SARS-CoV-1 in 2002 in China; and SARS-CoV-2 in 2019–2020 in China. In the 1990s, it was already possible to identify and characterise the infectious agents behind new viral diseases within a few months. Today, a few days is sufficient, given that international networks of research teams collaborate to tackle such challenges. Sometimes, "abnormal" illness may first be identified in animals, as was the case for BSE in 1991 (see p. 105). One major sign of human-to-human transmission is when an infection moves from patients to health care workers. Health investigations are launched based on information passed along by professionals in the field and reports of "abnormal" disease clusters by hawk-eyed, specialised health officials, a process that involves alerting different stakeholders.

Work is underway on syndrome-based surveillance programmes, which would assess automatically recorded data analysed in real or near-real time and thus help better detect disease emergence events. For example, these efforts may focus on analysing mortality rates in emergency care facilities or consumption patterns of certain medications. Since the 2000s and especially since 2010, it has become possible to analyse billions of data points collected in real-time thanks to advances in informatics, machine learning, and artificial intelligence. These data are diverse, ranging from information collected by airline companies to the reports written by governmental health authorities. Because events must be detected without any knowledge of their origin, there is a need to distinguish between normal events (i.e., background noise) and abnormal events (i.e., potential instances of disease emergence). The ultimate objective of the above tools is to identify abnormal events earlier on than is possible when detection is based on observed disease clusters. One challenge is that syndrome-based surveillance relies on continuously collected data, which exhibit fluctuations as a result of normal dynamics. Therefore, events must be dramatically different for signals to stand out within the data.

Pathogen Identification

Historically, microorganisms were defined as pathogenic based on Koch and Hill's postulate (1890): they must be present in sick individuals but absent (or rarely present) in healthy individuals; they must be capable of being cultured; and they must cause disease if used to infect healthy individuals. As our depth of knowledge has grown, these criteria have been updated.

Thanks to high-throughput sequencing, we can fully characterise the nucleic acids in samples fairly quickly and inexpensively. When microorganisms can be cultured, it is essential to show that the potential pathogen is infectious in test samples and to study the microorganism's characteristics to develop precise diagnostic tools and targeted treatments. In this sense, high-throughput sequencing has the advantage of uncovering new microorganisms with the caveat that neither their pathogenicity nor zoonotic potential is known. Consequently, this technique is also used to catalogue potential pathogens. A 2018 study by the US-based Global Virome Project estimated that mammals and birds host 1.7 million unknown viruses distributed across 25 virus families. To estimate how many of these viruses could be zoonotic, researchers conducted an analysis taking into consideration the relationships between animal species and known viruses, the history of viral zoonoses, and patterns of virus emergence. Based on their assessment, 700,000 of these 1.7 million unknown viruses have the potential to infect humans. Please note that this work has purely estimated the potential to cause infection in humans, not the potential to result in disease emergence. Furthermore, this number is a rather rough ballpark figure, likely far too high or low. For example, since the discovery of SARS-CoV-1, hundreds of viruses have been identified in bats. Yet, SARS-CoV-2 has never been observed in any of the samples. It seems likely that SARS-CoV-2 resulted from two coronaviruses recombining in an intermediary host, whose identity was still being debated when this book was written. Indeed, even if researchers have identified the main traits of emerging zoonotic viruses (see sidebar p. 12), it remains unlikely that we will be able to predict the next viral zoonosis to go epidemic.

Research utilising experimental infections *in vitro* or *in vivo* (i.e., in laboratory animals) can explore a pathogen's adaptability and pathogenicity in different potential animal reservoirs or animal sources of transmission to humans. This type of work is essential because pathogenicity is determined by a microorganism's properties; its host's physiological state and reaction to infection; and the ambient microbial environment.

A combination of methods is needed to establish whether a given pathogen causes a certain disease. For example, to confirm that SARS-CoV-2 was behind the clinical symptoms of COVID-19, the following methods were used: genomic sequencing of samples obtained via bronchoalveolar lavage or throat swabs; viral isolation techniques; and pathogenicity testing, which verified that the virus was causing the observed clinical symptoms.

Identifying the Infection Source

Identifying the source of a zoonotic infection is often a long process, requiring expertise in several domains: molecular biology, epidemiology, ecology, the social sciences, and the humanities. Strong evidence exists in the form of genetic similarities in the pathogens found in humans *versus* potential animal reservoirs. However, such evidence is not always found. Moreover, it cannot stand alone as we must also clarify transmission dynamics: who is transmitting the pathogen to whom and under which circumstances? Causal links must be established based on epidemiological, medical, and experimental research, which may include modelling. Various studies examine the frequency and strength of any associations as well as the associations' chronological consistency and specific nature. Any potential sources of bias must also be explored.

For example, while the source of Ebola virus has been established (i.e., forest primates), the identity of the reservoir remains hypothetical (i.e., fruit bats). SARS-CoV-2 has a genome that is 96% identical to that of a virus found in Asian horseshoe bats (genus *Rhinolophus*). However, we still do not fully understand the relationship between these two viruses, nor do we have a grasp on when SARS-CoV-2 actually emerged, only that it was detected in humans for the first time in late 2019. This issue is also illustrated by hepatitis E virus, which infects an estimated 20 million people worldwide per year. Most cases arise as part of epidemics, which largely take place in low- and middle-income countries. However, human populations in industrialised countries carry antibodies specific to hepatitis E virus, suggesting the presence of animal reservoirs. In Japan, individuals fell ill after eating raw pork, providing evidence of direct transmission. Cases have also resulted from the consumption of undercooked wild boar meat. The results of several epidemiological studies therefore support that this virus is transmitted from animals.

ARE ZOONOSES BECOMING MORE FREQUENT?

When smallpox was eradicated in the 1970s, certain authorities within the medical world predicted the end of all microbial diseases. AIDS immediately arrived on the scene, hand in hand with a rise in antibiotic resistance. It was a painful reminder that public health could take an entirely different course, which it did across the world in the decades to come. Some previously unnoticed phenomena have become noticeable, as the human population has climbed rapidly in size and our ability to detect diseases has grown. Thus, are we actually witnessing an increasing number of zoonotic epidemics, as the media has been suggesting?

Different approaches have been developed to analyse patterns of zoonotic epidemics and their associated factors. One approach is to study the occurrence of zoonoses, using data in international databases. Another approach is to conduct meta-analyses, which evaluate the results of several scientific studies and can thus identify general trends and potential explanatory factors. Finally, targeted field or laboratory research can be used to test specific hypotheses. In all the above approaches, researchers try to account for confounding variables, including the effort invested in data collection or healthcare system quality, using metrics such as the estimated number of publications on the target topic, levels of healthcare funding, and country economy size.

As highlighted in the sidebar, the number of zoonosis epidemics has increased over time. We see the same dynamics for epidemics of human infectious diseases in general (i.e., zoonotic and non-zoonotic). Zoonosis diversity has grown in tandem with epidemic frequency. As a consequence, we are experiencing more epidemics representing a broader range of zoonoses. Zoonosis emergence is mainly being driven by a complex set of factors:

changing interactions at the interface between wild vertebrates, domestic vertebrates, and human beings under conditions of rapidly shifting land use (i.e., agricultural intensification, urbanisation, and deforestation).

CHANGES IN ZOONOSIS EPIDEMIC FREQUENCY BASED ON THE GIDEON DATABASE

To explore the question raised above, we will use the information available in the Global Infectious Diseases and Epidemiology Network (GIDEON) database, currently the most comprehensive source for data on human infectious and parasitic diseases. The information contained in the database has been verified by experts. It brings together WHO data, scientific findings published in international journals, and historical data on epidemics dating back several centuries. It also employs the WHO's definition of an epidemic, which focuses on established causality and/or chains of transmission rather than on a threshold number of cases. However, like any data source, it has its particular biases. Notably, different countries may vary in how well their disease surveillance programmes pick up on or report certain epidemics. These differences are due to a multitude of factors. In addition, research is greatly lacking for many of the so-called neglected tropical diseases.

Drawing upon the GIDEON database, we plotted the number of reported zoonosis epidemics over time. There is a clear increase from 1960 onwards, with two major episodes corresponding to H1N1 influenza in 2009, caused by the A(H1N1)pdm09 virus, and COVID-19 in 2020, caused by SARS-CoV-2 (see Figure 14). This general pattern aside, there has been a dip in epidemic frequency over the last two decades. Is this a short-term trend or the beginning of an epidemiological transition? Only the future will tell.

In the GIDEON database, diseases are classified according to the number and type of organisms involved in maintaining pathogen transmission. Because we are interested in zoonoses, we have removed strictly human diseases as well as diseases that are only associated with arthropods (e.g., malaria, with the exception of the types caused by *P. knowlesi* and *P. cynomolgi*) or molluscs (e.g., schistosomiases that do not utilise a major vertebrate reservoir, like *Schistosoma mansoni*). We also excluded cases of antibiotic resistance because it is often difficult to objectively identify the zoonotic

.../...

...⁄...

origin of resistance. There was some debate about whether or not to include diseases such as dengue, chikungunya, or Zika fever. While these diseases are caused by viruses that emerged from non-human primates, they are now essentially transmitted among humans outside the areas in which they emerged. However, we decided to treat these diseases as zoonotic in our analysis given there is no evidence that non-human primates no longer contribute to local virus transmission. COVID-19 was also included in our list.

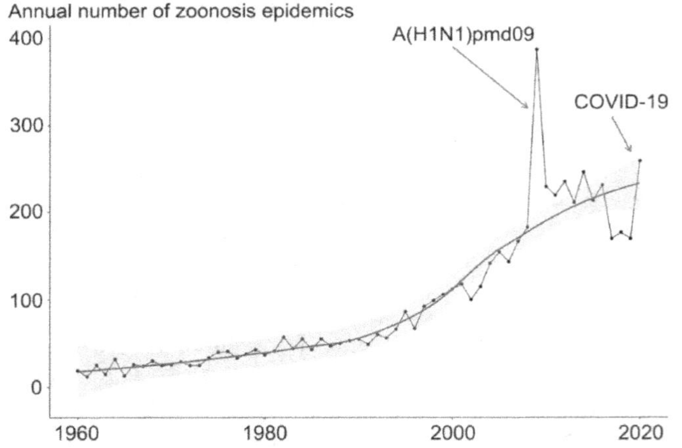

Figure 14. Number of zoonosis epidemics over time.
The data selection criteria are described in the text. Indicated are the years for which data were available in the GIDEON database (through June 2020). The solid line is the smoothed mean, and the grey envelope is the confidence interval of the smoothed mean (© Serge Morand).

It is also enlightening to examine the geographical patterns associated with zoonosis emergence. Numerous modelling studies have sought to identify emergence hotspots by focusing on interfaces between humans and other animals. The higher-quality models include interactions between various hazard-related metrics, such as pathogen diversity, biodiversity, and farm animal densities. They also incorporate indicators that convey the likelihood of human exposure, such as human population densities or levels of habitat destruction, and indicators of vulnerability, such as the degree of healthcare funding.

As previously mentioned, one challenge is that our understanding of biodiversity is biased by our relative degree of interest. That said, zoonoses most often emerge in Southeast Asia, India, Europe, parts of China, parts of Central and South America, and tropical zones in Africa. The density of human populations in these places likely plays an influential role, especially in countries like India or China. The economy also has an important part to play in the most industrialised countries. Because such countries are part of a broader economic web, they are at greater risk of experiencing pandemics. Furthermore, it is in these countries that surveillance programmes and detection efforts take on the most importance.

Indeed, in addition to becoming more frequent, zoonosis epidemics have also gone more global since the 1970s. From that point on, epidemics tended to display broader, worldwide distribution patterns. This globalisation of zoonoses is linked to the greater movement of people and live animals. For example, the annual number of airline passengers grew from 330 million in 1970 to over 4 billion in 2019. Live cattle are also moving around at far higher levels. Worldwide, estimated transportation-related expenses rose from US$2 billion in the 1970s to more than US$18 billion in 2017.

ROLE OF BIODIVERSITY

Since the Neolithic, there have been dramatic shifts in the relative biomass contributions of wild vertebrates, domestic vertebrates, and humans (see Figure 15). Furthermore, it is apparent that the diversity of zoonotic pathogens is positively correlated with the diversity of available hosts: all microorganisms, pathogens and non-pathogens alike, are indeed an integral part of biodiversity.

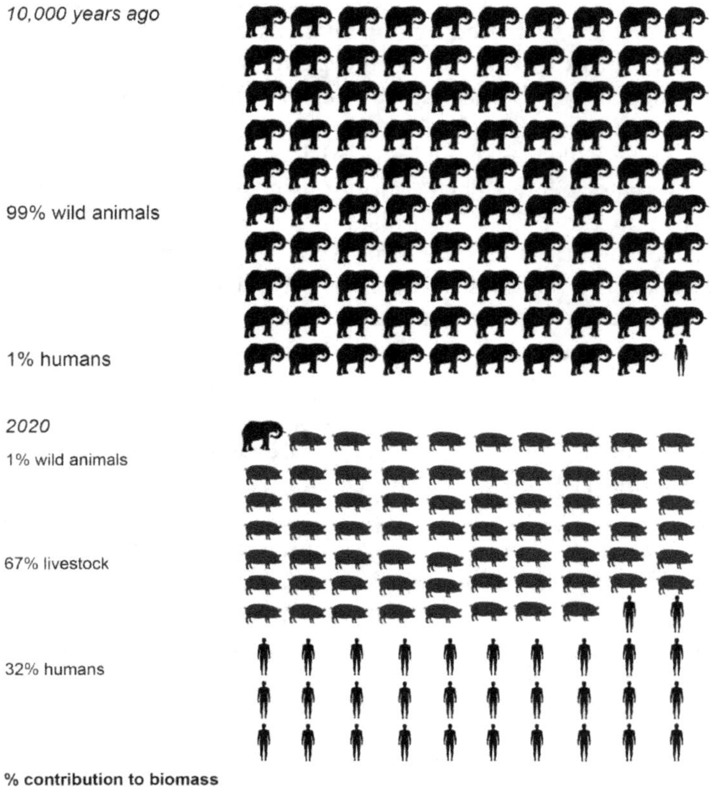

Figure 15. Relative contributions of different terrestrial groups to vertebrate biomass between the Neolithic and the present.

Adapted from Smil, 2011 (https://doi.org/10.1111/j.1728-4457.2011.00450.x).

Two seemingly paradoxical hypotheses have emerged from research focused on the relationship between biodiversity and infectious or zoonotic diseases. The "diversity begets diversity" hypothesis posits that any increase in host diversity is positively correlated with overall pathogen diversity. This relationship is what we observed in our exploration of the association between zoonosis frequency and animal species richness across countries (see Figure 16). The "dilution effect" hypothesis is rooted in ideas about predator-prey relationships, notably that an increase

BIODIVERSITY AND ZOONOSES

We explored the relationship between biodiversity and zoonosis emergence by looking at maps of IUCN data from 2019 and maps of GIDEON data showing zoonosis frequency from 1960 to 2019. To compare patterns across countries, we corrected epidemic frequency based on the number of known diseases per country. The number of endangered species per country was also corrected based on confounding variables such as the abundance of known animal species per country.

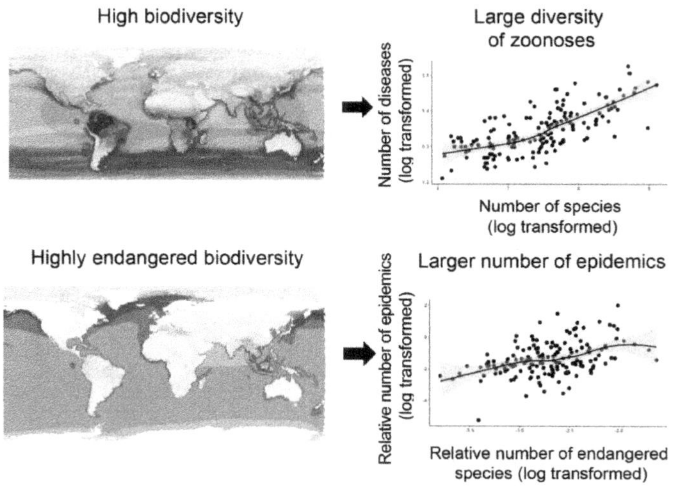

Figure 16. Relationships between biodiversity and zoonosis frequency per country (top panel) and between the number of endangered species and the frequency of zoonosis epidemics (bottom panel).

Top map: Species richness—terrestrial and marine mammals (the dark areas are mammalian biodiversity hotspots); **Bottom map:** Species richness—endangered terrestrial and marine mammals (the only dark area is in Southeast Asia); **Top figure:** Relationship between zoonosis frequency (GIDEON data) and animal species richness (IUCN data) for different countries across the globe; **Bottom figure:** Relationship between the relative frequency of zoonosis epidemics (GIDEON data) and the relative richness of animal species (IUCN data) for different countries across the globe. From Morand and Lajaunie, 2017 (https://www.sciencedirect.com/book/9781785481154/biodiversity-and-health#book-description).

in prey number means that any given individual becomes less likely to face predation. When applied to infectious diseases, and primarily to Lyme disease, the dilution effect hypothesis posits that high host diversity should "dilute" the epidemiological role played by the main reservoir species. In other words, greater biodiversity should lead to lower levels of transmission as pathogens rarely encounter their natural host species. Thus, host richness and diversity should have a protective function when it comes to pathogen diffusion. The flip side of the dilution effect hypothesis is that declines in biodiversity could theoretically promote pathogen spread. Several meta-analyses have shown that various diseases affecting humans, wildlife, trees, and other plants display evidence of a dilution effect. The occurrence of a dilution effect has been demonstrated for many diseases at multiple scales, from local to global.

Mechanistically, the dilution effect results because high biodiversity translates into greater food web diversity, and, in particular, the presence of predators that regulate certain reservoir and vector populations. When reservoir species are no longer regulated by predators, have no competitors, or are highly adapted to anthropogenic habitats (e.g., fragmented habitats), then reservoir populations expand, facilitating the transmission of the agents they host and thus increasing infection risks for other animals, including humans (see Figure 17).

Ecosystem services are ecosystem functions that contribute to societal needs and that improve individual and collective well-being. There are four general types of services: provisioning, regulating, cultural, and supporting. The dilution effect can mechanistically contribute to infectious disease regulation. Another facet of this service is that humans are exposed to a greater variety of antigens when biodiversity is greater (see p. 112). Biodiversity also supplies natural compounds that can be used to fight pathogens and helps limit levels of pollution, which has harmful effects on immune function. However, although many studies have examined ecosystem services related to climate regulation or water purification, much more research should explore how well-functioning ecosystems could help regulate infectious diseases. A 2015 meta-analysis assessed how ecosystem disservices (i.e., biodiversity losses and gains) could have negative effects on human or animal health, such as spurring allergies or promoting the spread of pathogen vectors. Particular attention was paid to urban and agricultural settings.

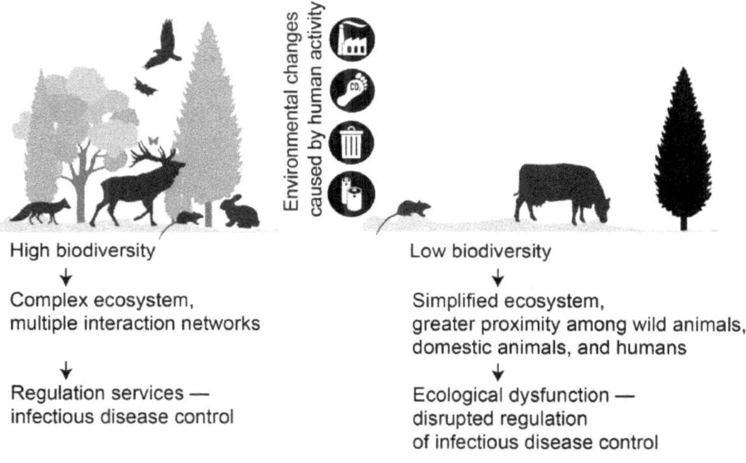

Figure 17. Biodiversity and its regulatory services in relation to infectious diseases.

ROLE OF FARM ANIMALS AND PETS

When we looked at the relationship between domestication/commensalism and pathogen sharing, we found that animals with longer shared histories with humans had more pathogens in common with humans and other domestic/commensal animals (see p. 39). There are three key consequences of this relationship that are worth noting:

- It takes time for new zoonoses to establish themselves in humans and synanthropic animals.
- There is no evidence that domesticated animals are done passing along zoonoses.
- Any new farm animal species is likely to contribute to the circulating network of infectious agents, either as a host or a donor.

A 2019 study explored the network of associations between mammals (724 species, including 21 domestic animals and humans) and viruses (1,785 DNA and RNA viruses, both zoonotic and human specific). The authors found that domesticated mammals occupied highly central positions, thus serving as epidemiological bridges between wildlife and humans. In particular, ungulates are responsible for sharing large numbers of both DNA and RNA viruses, while carnivores tend to only share the latter. It should also be noted that bats share a large number of RNA viruses (but see sidebar p. 70). Similar results were obtained from a study investigating the network of associations between humans and other animals for shared intracellular bacteria of the genus *Rickettsia*. Some species in this genus are vectored by ticks, and many have domestic animals as reservoirs.

Over recent decades, there has been a dramatic increase in the size of farm animal populations. Worldwide, between the 1960s and the present, the number of livestock has climbed from 1 billion to 1.6 billion head, while the number of chickens has soared from 4 billion to 30 billion, fast approaching the estimated 50 billion members of wild bird populations. Such population growth influences the dynamics of zoonoses in several ways. First, farms can act as pathogen incubators. Industrial farms implement stringent biosecurity measures to prevent infectious diseases (see p. 123), but these measures do not always meet global standards. Pathogens are likely to spread extremely rapidly upon arrival if farm conditions include high animal densities, stressful rearing conditions, and animals with low genetic diversity. Livestock farming also plays an indirect role in zoonosis dynamics via the landscape modifications it induces. In many countries, habitats composed of small natural areas and/or farmed plots have been replaced by large plots containing fast-growing monocultures (e.g., of corn or soybeans) that are used to feed intensively raised livestock. In such farming systems, animals do not experience natural physiological or dietary conditions. For example, cattle are no longer entirely grass fed. These landscape transformations have highly detrimental effects on biodiversity and play a role in zoonosis emergence, as we will discuss below. Finally, the intensification of livestock farming and the industrialisation of agricultural systems have resulted in the massive deployment of numerous biocides, particularly antibiotics. The resulting resistance genes can be transferred to bacteria hosted by humans (see p. 82). Furthermore, farmers also employ a range of inputs and chemical compounds that damage ecosystems and weaken organismal defences against infections. The worldwide expansion of industrial livestock farming is hazardous for human health, animal health, and ecosystem health.

WILDLIFE TRAFFICKING

A 2019 study estimated that, overall, international trade in exotic pets (amphibians, birds, reptiles, and mammals) resulted in the legal exportation of more than 11 million live animals between 2012 and 2016. Represented among these animals were 1,316 different species from 189 countries. Most often, countries of the Global South were the exporters, while countries of the Global North were the importers. These figures do not include those for the illegal exotic pet trade, for which statistics are far more complicated to obtain.

In some countries, these animals are sold in markets under shocking conditions, namely crammed together in large piles of cages. It is hard to avoid thinking of these situations as other than an enormous natural experiment exploring microorganismal exchanges among species, including humans. Furthermore, in tropical regions of the world, many people use local wild species as their primary source of animal protein; these animals are obtained via hunting or from commercial sources. While it used to be that most consumption occurred locally, such meat is now massively exported. This trend stems from increasing migration, the development of transportation networks, and economic shifts. The amounts involved are hard to quantify because bush meat importation is officially banned for health and safety reasons. That said, it is frequently estimated that, in planes coming from certain tropical regions, passengers have an average of 1 kg of bush meat in their hand luggage (i.e., 200 to 300 kg per flight).

LAND USE CHANGE

Natural ecosystems, especially forests, are experiencing increasing pressure as humans extract resources and convert landscapes. Intertropical zones harbour high levels of biodiversity. Their deforestation results in new contacts between wild species, domestic species, and humans. For example, in Asia, as irrigated agricultural areas have expanded, so have the breeding areas of *Culex* mosquitoes, which vector Japanese encephalitis virus. Wild birds are the reservoirs for this virus, which has established a secondary cycle in pigs, the source responsible for human infections.

CHANGES IN FOREST COVER

The GIDEON database can also be used to explore the relationship between zoonosis epidemics and changes in forest cover. The latter have namely arisen from the increase in commercial palm oil plantations, based on information from the FAOSTAT database. Over the period from 1990 to 2016, it is clear that higher levels of deforestation are associated with more frequent zoonosis epidemics (see Figure 18). These results are consistent with previous findings. Notably, research conducted since the mid-2010s has shown that land use changes, including the conversion of forests, favour populations of zoonotic reservoirs and, consequently, boost the risk of zoonoses. Additional factors associated with deforestation may also promote zoonosis epidemics, including increasing levels of anthropogenic activities, declines in biodiversity, especially that of large predators, and disruptions in community functioning.

Furthermore, the greater the surface area dedicated to palm oil plantations, the more frequently zoonosis epidemics were seen. Studies have already underscored that the expansion of palm oil plantations is negatively affecting biodiversity, particularly in Southeast Asia and South America. This trend has been illustrated by the emergence of Nipah virus (see p. 69).

A 2019 meta-analysis using data from Southeast Asia showed that the expansion of palm oil monocultures increased the likelihood of zoonoses, such as leptospirosis, rickettsial diseases, and malaria caused by *P. knowlesi*, for which the reservoirs are macaques. In Colombia, kissing bug populations thrive on palm oil plantations. These insects vector the protozoan *Trypanosoma cruzi*, which causes Chagas disease and has multiple reservoir hosts.

.../...

.../...

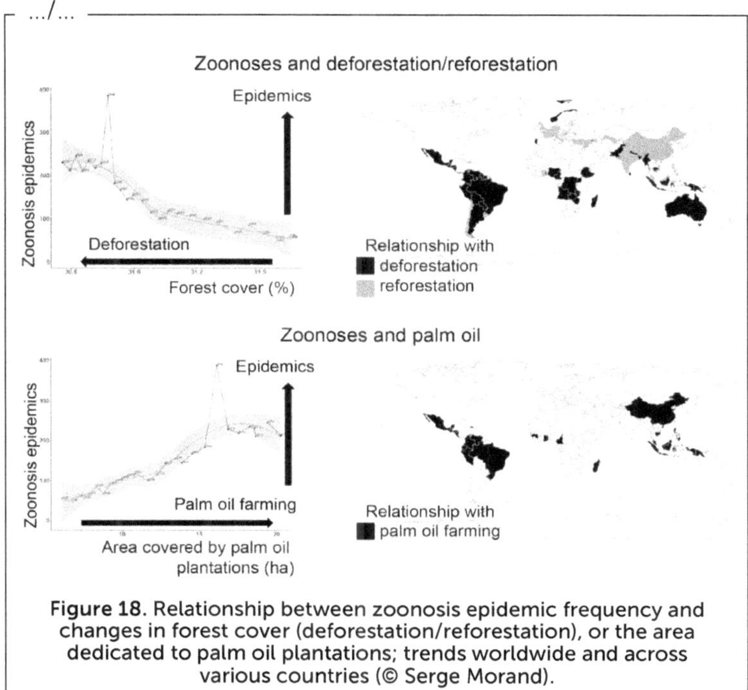

Figure 18. Relationship between zoonosis epidemic frequency and changes in forest cover (deforestation/reforestation), or the area dedicated to palm oil plantations; trends worldwide and across various countries (© Serge Morand).

France has witnessed an increase in its amount of forested land, which has been climbing by 0.7% per year since 1990 and which reached more than 16 million hectares in 2020 (i.e., 31% of the country's surface area). While a similar trend has been seen in Europe, the rest of the world is experiencing high levels of deforestation. At present, two-fifths of Europe is covered by forests and woodlands. Between 1990 and 2015, approximately 90,000 square kilometres were reforested. However, such increases can result from two dramatically different situations: 1) an expansion of non-natural forests planted by humans, which generally have low biodiversity, and 2) an expansion of naturally reforested land, namely abandoned agricultural fields or grazed grasslands. Increases in forested land may be linked with greater epidemic frequencies, particularly in non-tropical countries with low to moderate levels of forest cover. For example, Italy has witnessed a resurgence in the incidence of tick-borne encephalitis

159

that is tied to increased levels of natural reforestation, which has boosted the numbers of the small mammals that serve as virus reservoirs. Similarly, tick-borne zoonotic diseases are on the rise in the US because of the favourable ecological conditions created by reforestation, burgeoning deer populations unregulated by large predators, and the diverse human activities that take place in forests. Furthermore, epidemic frequencies are climbing in tropical countries engaging in extensive reforestation, such as China, Malaysia, the Philippines, and India. This trend is mainly associated with the creation of single-species plantations that particularly foster populations of synanthropic wild species (e.g., rats, mosquitoes, and other arthropods).

CLIMATE CHANGE

In 2022, there is broad scientific consensus that climate change is a reality and that humans are responsible. According to the IPCC's 2021 report, the planet is currently +1.11°C warmer than it was in pre-industrial times. Furthermore, we are witnessing other impacts, such as melting ice caps and rising sea levels, as well higher-frequency and greater-intensity extreme weather events. These changes are affecting all life on Earth, including the planet's myriad biological interactions. The ecology of zoonoses has not been spared. That said, it is not an easy task to pinpoint the exact effects of climate change. There are several reasons. First, it is challenging to tease apart the forces in operation given the diverse and complex relationships that exist between climatic conditions and zoonotic cycles. Second, we are witnessing other major global changes that also affect zoonosis dynamics, including changes in land use, the intensification of livestock farming, socioeconomic transitions, and dramatic population shifts.

Climate change can affect the ecology of zoonoses by altering the range of vectors or hosts. For example, it appears that the tick *Ixodes ricinus*, the vector of Lyme disease in Europe, has expanded its range northward in Scandinavia (i.e., beyond latitude 60° N). It is now also found at higher altitudes. Another illustration is the vectors that were introduced into new areas via long-distance travel by people, farm animals, or migrating birds and that have managed to establish themselves permanently. Such has been seen for the tick *Hyalomma marginatum* in southern Europe (see p. 92) and the tiger mosquito, *Aedes albopictus*, in Europe more generally. Climate change also affects the active season and development of vectors and hosts. Taking the example of *I. ricinus*, we are observing increasing numbers of ticks that are active in the winter. Finally, climatic conditions can directly affect pathogen survival and development. A 2017 study found that 99 of 157 zoonotic pathogens in Europe (63%) were sensitive to climatic conditions.

In addition, climate change may increase human susceptibility to diseases in general, notably by provoking more frequent heat waves.

ZOONOTIC RISKS OF MELTING PERMAFROST

Permafrost is a soil type that remains at temperatures below 0°C for at least two consecutive years. Around 20% of the planet is covered by permafrost. Concerns have been raised over the permafrost melting because of the climatic impacts: this soil type contains nearly 1,700 billion tonnes of greenhouse gases, which is about twice the amount of carbon dioxide already in the atmosphere. The permafrost also harbours bacteria and viruses, including some that are quite old (> 10,000 years). In samples of frozen animal skin and fur, scientists have uncovered new viruses, including giant viruses that exceed 0.5 μm in diameter. In the summer of 2016, a child in Siberia died from anthrax, also known as the "Siberian plague" (see sidebar p. 45). The child was probably infected by bacteria on the carcass of a reindeer that had died several decades ago, which thawed out and ended up contaminating present-day reindeer herds. Back in the 19th century, there were already references to the infection of hundreds of thousands of cervids. Between 1985 and 2008, around ten thousand bovines and reindeer died of anthrax, and most of their corpses were buried in the permafrost. While the bacteria responsible for anthrax are highly resistant to cold outdoor conditions, questions remain about the viability of the viruses in the permafrost that have experienced freezing and thawing cycles.

CONCLUSIONS: WHAT COMES NEXT?

This book has clearly illustrated that zoonoses are diseases whose infectious agents are transmitted between one or more vertebrate species and our own species, *Homo sapiens*. These diseases have always been with us. Some zoonotic pathogens, such as those that cause tuberculosis, have even been known to move back and forth between humans and other animal species. For other zoonotic pathogens, humans represent an epidemiological dead end. Yet others have adapted to allow transmission among humans. The geographical distribution and frequency of zoonoses is moulded by our evolutionary history, our use of the planet's available space, and our interactions with the living world.

We can better prevent zoonoses via individual and collective actions that are based on a sound knowledge of pathogen sources and transmission routes. Zoonotic risks should not result in us completely cutting ties with animals. When pets are kept healthy, under conditions that ensure their safety and welfare, they pose minimal risks to humans with properly functioning immune systems. However, the fact that zoonoses are emerging from wild fauna or resulting from our growing exploitation of planetary resources is another sign among many that we must rethink our relationship with the living world.

The Anthropocene presents us with a reality that cannot be denied, even if it remains to be formally recognised as a geological era. Some people may seek reassurance in the belief that all our problems can be solved exclusively through scientific and technological solutions, notably via an increase in the physical and mental capacities of humans. This hope is in vain. As science philosopher Vinciane Despret has noted, "A tiny virus has managed to block the entire economy, which the climate emergency has failed to do!" In turn, writer Sylvain Tesson has commented that the emergence of SARS-CoV-2 "displays all the hallmarks of a modern phenomenon: it has been rapid, massive, global, and uncontrollable". This crisis has brutally reminded us that we too are living creatures. We are mortal. We share our planet with other forms of life. We are all connected, even though we may occur on opposite sides of the world. The Anthropocene sounds the death knell of the idea that humans exist outside of nature. It is an undeniable fact that humans cannot be separated from their environmental surroundings. Humans are only one of the many organisms to be found within the "ecosphere" that is Earth. Our lives depend upon the planet and all its non-human species. For our species to persist as long as possible, we must strike a new balance with the rest of the living world by better protecting its ecosystems. The Earth does not belong to us. On the contrary, we belong to the Earth.

The oldest part of our brains, the one already present in our primate ancestors millions of years ago, is programmed for survival in the world's forests and savannas. It is not equipped to temper our actions and plan ahead in a world that presents us with abundance, albeit abundance that remains reserved for a privileged few. Yet we must espouse moderation if we are to change course and rethink the damage we are inflicting on the planet. The time has come to unleash the power of our imaginations, as suggested by Rob Hopkins, who initiated the Transition Town Movement. We must let our brains feel the world again and wean them off of the one hundred and one objects that clutter up our lives and pollute the oceans. We need to nourish ourselves with knowledge and meaning. Humans must tackle the environmental challenge that is sustainable development. We must build new models of local and international governance. We must develop public and economic policies that promote sustainability at the planetary level. It is essential to recognise that sustainable development is incompatible with constant growth. We need to rethink our attachment to the notion of growth. We face the responsibility of building a world in which preserving life is at the centre of our concerns, not money, the financial markets, or the creation of monetary value. In this world, there would be a place for all humans, including those who are currently suffering, as well as for all non-humans, including animals and their microbes. We must question our current convictions and reinvent our current practices. We are capable of developing relationships with other living beings and with animals in particular. We share with them a universe of emotions and sensations that touch our five senses. Their cognitive abilities are at the heart of our strong bonds with them, as evidenced by the release of oxytocin, a social bonding hormone, when a dog's owner glances at their animal companion. Our health is inextricably linked to that of our ecosystems. Let's take care of them and us.

LEARN MORE

ONLINE RESOURCES

Glossaries

Medical terminology: https://www.msdmanuals.com/home

Websites for Health Organisations

Animal Health Epidemiology Surveillance Platform (ESA): https://www.plateforme-esa.fr/ (in French)

European Centre for Disease Prevention and Control (ECDC): https://www.ecdc.europa.eu/en/zoonoses

European Food and Safety Authority (EFSA): https://www.efsa.europa.eu/en

World Health Organisation (WHO): https://www.who.int/news-room/fact-sheets/detail/zoonoses

World Organisation for Animal Health (WOAH): https://www.woah.org/en/home/

Examples of One Health network or initiative

From One Health to Ecohealth: https://www.iddri.org/en/publications-and-events/issue-brief/one-health-ecohealth-mapping-incomplete-integration-human

International Center for Well Being: https://onehealthplatform.com/

Network of evaluation of One Health: http://neoh.onehealthglobal.net/

One Health High Level Expert Panel: https://www.who.int/groups/one-health-high-level-expert-panel

One Sustainable Health: http://www.onesustainablehealth.org/fr/

Preventing zoonotic diseases emergence: https://prezode.org/prezode_fre/

SUGGESTIONS OF REPORTS AND BOOKS

Reports

Report of the Harvard Global Health Institute. Alimi *et al. Report of the scientific task force on preventing pandemics.* 2021, 36 pages. https://www.hsph.harvard.edu/c-change/news/PreventingPandemicsResearch/

Report of the IPBES (Intergovernmental Science-Policy Platform on Biodiversity and Ecosystem Services). *Workshop report on biodiversity and pandemics.* 2020, 96 pages. https://ipbes.net/pandemics

Report of the IUCN (International Union for Conservation of Nature). Kock & Caceres-Escobar. *Situation analysis on the roles and risks of wildlife in the emergence of human infectious diseases.* 2021, 111 pages. https://portals.iucn.org/library/node/49880

Books

Gavier-Widen Dolorés, Meredith Anna, Duff J. Paul. *Infectious diseases of wild mammals and birds in Europe.* Wiley-Blackwell 2012. 568 pages.

Sing Andrea. *Zoonose-infection affecting humans and animals. Focus on public health aspects.* Springer 2015. 1143 pages

Waltner-Toes David. *On Pandemics: Deadly diseases from bubonic plague to coronavirus.* Greystone books 2020. 248 pages

Zinsstag Jakob, Schelling Esther, Crump Lisa, Whittaker Maxine, Marcel Tanner, Stephen Craig. *One Health. The theory and practice of integrated health approaches* (2nd edition). CAB International 2020. 464 pages

By the same authors

Gardon S., Gautier A., Le Naour G., Morand S., 2022. *Sortir des crises. One Health en pratiques.* Éditions Quæ.

Morand S., 2016. *La prochaine peste. Une histoire globale des sociétés et de leurs épidémies.* Éditions Fayard.

Morand S., 2020. *L'homme, la faune sauvage et la peste.* Éditions Fayard.

Morand S., Figuié M., 2017. *Emergence of infectious diseases. Risks and issues for societies.* Éditions Quæ.

Morand S., Lajaunie C., 2017. *Biodiversity and Health. Linking Life, Ecosystems and Societies.* Elsevier, London.

Morand S., Moutou F., Richomme C., 2014. *Faune sauvage, biodiversité et santé, quels défis?* Éditions Quæ.

Moutou F., 2020. *Des épidémies, des animaux et des hommes.* Editions Le Pommier.

Moutou F., 2021. *Adopte un virus.com: Quand les microbes passent de l'animal à l'homme.* Delachaux et Niestlé.

Acknowledgements

The authors extend their warmest thanks to Valérie Mary and Véronique Véto for their support along the path to this book's publication. We are grateful to Jessica Pearce-Duvet for translating the book into English.

Translation: Jessica Pearce-Duvet

Editorial Operations for the English Version: Alana Claudin

Layout: L EliLoCom

Imprimé pour vous par Books on Demand (Allemagne)